tommy walsh
BATHROOMDIY

tommy walsh
BATHROOMDIY

Collins

To my wife Marie and my kids, Charlotte, Natalie and Jonjo, for putting up with my manic lifestyle, and the fact that on the rare occasions that I am home, I'm often secreted away in my study writing.

First published in 2004 by Collins an imprint
of HarperCollins*Publishers*,
77-85 Fulham Palace Road, London, W6 8JB

The Collins website address is: www.**collins**.co.uk

Text copyright © Tommy Walsh
Photography, artworks and design © HarperCollins*Publishers*

Designed and produced by Airedale Publishing Ltd
Art Director: Ruth Prentice
PA to Art Director: Amanda Jensen
Editor: Jackie Matthews
Designer: Hannah Attwell
Assistants: Andres Arteaga, Anthony Mellor, Neal Kirby
DTP: Max Newton
Tommy Walsh photographs: David Murphy
Other photographs: Sarah Cuttle, Mike Newton, David Murphy
Artworks: David Ashby
Consultant: John McGowan
Index: Emma Callery

For HarperCollins
Senior Managing Editor: Angela Newton
Editor: Alastair Laing
Design Manager: Luke Griffin
Editorial Assistant: Lisa John
Production Controller: Chris Gurney

A CIP catalogue record for this book is available
from the British Library

ISBN: 0007156898
Colour reproduction: Colourscan
Printed and bound: Lego, Italy

684

contents

introduction

Bathrooms need to be well planned and functional, but they must also be a pleasure to use. The bathroom is certainly a special place for me, and I spend a considerable amount of time in mine; an invigorating shower in the morning sets me up for the day, and, when time allows, a long, hot soak in my huge, cast-iron, rolltop bath, with wonderfully scented oils, soothes my aching joints. And when I'm not indulging myself, my family is making maximum use of it!

A completely new, professionally installed bathroom may seem to be the ideal when a change is required, but may cost too much for the family budget and is often not necessary. Much can be done to create the right atmosphere in your existing bathroom without ripping the whole lot out. Indeed, all that may be required is to tidy up the tiling, swap the bath taps for a mixer, or replace a flimsy plastic bath panel for a more attractive one. If you are more ambitious, you could install a shower, or maybe even a wet room with underfloor heating.

There's a lot you can do yourself to achieve your dream bathroom if you are prepared to put in the hard work. If you feel that you're not sufficiently DIY-competent to tackle some of the more complicated jobs, however, you can still save yourself a lot of cash by doing all the stripping out and preparation before calling in the professionals. Even if you decide to leave all the work to plumbers and builders, it's a good idea to gen up on a project before you spend a considerable amount of money on it, so that you can plan the project efficiently and really know what you are talking about. Whatever you decide to do, my intention with this book is to guide you through all the stages.

Enjoy your new bathroom!

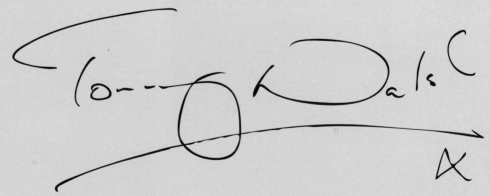

tools

Depending on your bathroom project, you may require a range of plumbing and tiling tools in addition to your general DIY toolkit. Write a list of the tools you need for the work, including any plastering, carpentry and decorating, then ring around local tool shops to check prices before you buy. Don't be afraid to ask for a deal; given the opportunity, specialist tool shops may be able to offer a better price than the DIY sheds, and may give more expert advice. Buy good quality tools and look after them – they should last a lifetime.

Listed here is a collection of tools that you may require. Before use, always read the safety information provided with a tool, and practise on scrap material before using it for the project.

LARGE TOOLS & DRILLS

always remember to read the safety information provided BEFORE use and practise before using for the first time

1 toolbox
2 14 volt cordless drill
3 18 volt cordless drill
4 cordless hammer drill
5 small cordless drill
6 drillbit selection
7 corded jigsaw
8 radial arm mitre saw (right)

GENERAL TOOLS

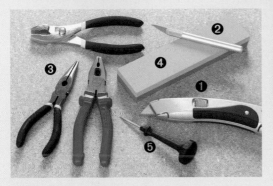

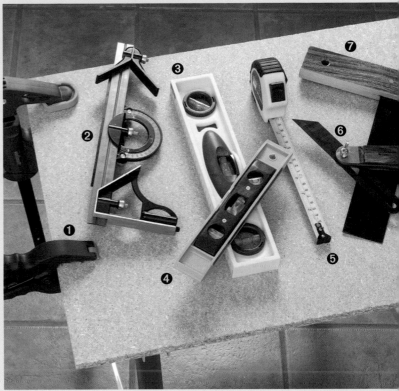

1 large craft knife
2 small craft knife
3 pliers & pincers
4 oilstone
5 bradawl

1 clamps
2 adjustable square
3 spirit level
4 pocket spirit level
5 tape measure
6 sliding bevel
7 combination square

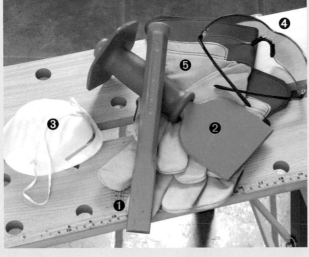

1 cold chisel
2 bolster chisel
3 dust mask
4 safety glasses
5 gloves

safety glasses and a mask are essential for any task that could produce flying pieces

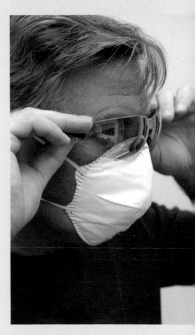

GENERAL TOOLS

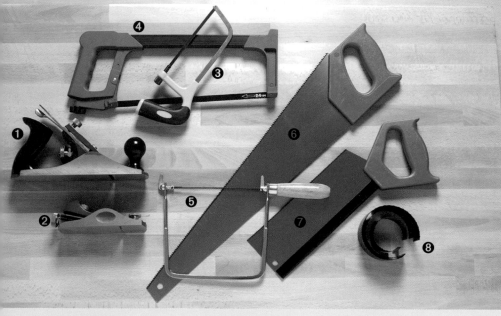

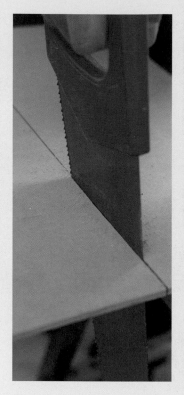

1 wood plane	4 large hacksaw	7 tenon saw
2 block plane	5 coping saw	8 set of hole saws
3 hack saw (junior)	6 hand saw	

1 claw hammer
2 chisel selection
3 screwdriver selection
4 rubber mallet
5 pin hammer

ELECTRICAL TOOLS

1 fuses
2 electric screwdrivers
3 tester screwdriver

4 long nose pliers
5 wire strippers
6 cable cutters

7 wire cutters
8 electrical tester
(above right)

PLUMBING TOOLS

1   adjustable spanners
2   gas torch & fire-proof mat
3   plier wrench
4   copper pipe cutter
5   PTFE tape
6   plastic pipe cutter
7   pipe repair tape
8   pipe repair clamp
9   hydraulic pump
10 sink plunger

DECORATING TOOLS

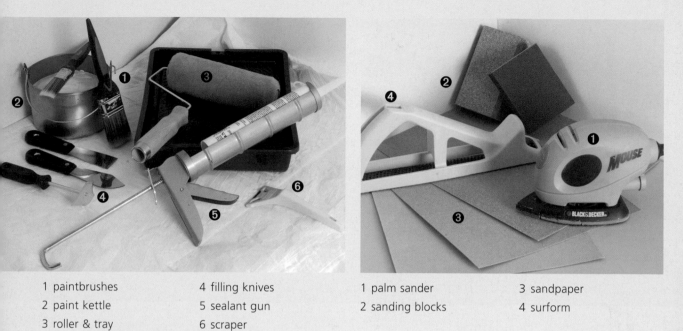

1 paintbrushes
2 paint kettle
3 roller & tray

4 filling knives
5 sealant gun
6 scraper

1 palm sander
2 sanding blocks

3 sandpaper
4 surform

TILING TOOLS

1 powered tile saw
2 tile cutting jig
3 grout shaper
4 grout remover
5 sponge
6 tile saw
7 rubber grout float
8 notched adhesive trowels
9 tile spacers
10 tile nibbler

bathroom basics

Where on earth do I connect these pipes? It makes me smile because I've been there! Firstly I'm going to explain the basics to you, so that when you have the feel for it, your chosen task won't be so daunting!

plumbing

In order to understand plumbing and how it works, it's essential to know where and how the water enters and exits your property, and what happens to it while it's in there. The mains water is supplied by the local water supplier to a stopcock either just inside or just outside your property boundary, via a mains stopcock. Any problems on the supplier's side of the stopcock are their responsibility, while any problems on your side of the stopcock are your responsibility.

DOMESTIC WATER SYSTEMS

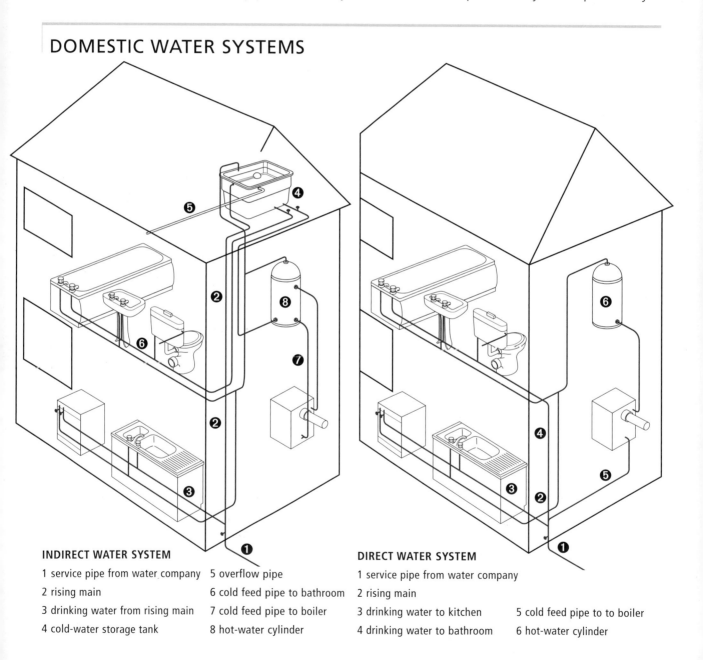

INDIRECT WATER SYSTEM

1 service pipe from water company
2 rising main
3 drinking water from rising main
4 cold-water storage tank
5 overflow pipe
6 cold feed pipe to bathroom
7 cold feed pipe to boiler
8 hot-water cylinder

DIRECT WATER SYSTEM

1 service pipe from water company
2 rising main
3 drinking water to kitchen
4 drinking water to bathroom
5 cold feed pipe to to boiler
6 hot-water cylinder

There are basically two types of domestic water systems, indirect and direct. Indirect is a mains-fed, stored-water system that supplies the drinking water then runs into a water tank in the loft which feeds all other water outlets.

Direct is also a mains-fed system but it feeds all the taps with mains water pressure and provides drinking water from any cold-water tap in the house. This is a great system, especially if you're contemplating a loft extension, as it does away with water storage tanks. It's also very handy if you happen to get thirsty in the night – you don't have to go all the way to the kitchen for a drink, just pop into the bathroom. The other advantages are that it's cheaper to install than the indirect system and you do not have to worry about the possibility of a water tank in the loft freezing and then flooding the house. A direct system requires a pressurized, unvented cylinder to store the hot water. However, for this system to work successfully you need good (strong) mains water pressure.

indirect water systems

The most common household system is the indirect stored-water system. This works from a mains supply from the company stopcock outside, which enters the house underground and usually surfaces near the kitchen sink. It supplies fresh drinking water under mains pressure, then travels via rising-main pipework to a large water tank in the loft space. This tank basically supplies all your water requirements other than fresh drinking water: usually a washbasin, bath, toilet and shower. The feed for the hot water comes via the storage tank to the boiler or heating cylinder. The water level within the tank is controlled by a floating ball valve which automatically shuts off the water supply at the pre-set level.

An overflow pipe fitted to the tank prevents flooding as a result of ball valve failure. Water pouring into the garden from the overflow pipe will quickly alert you to a ball valve problem. This is easily solved, usually by replacing a simple washer on the valve in the storage tank.

direct water systems

Here, the water supply comes in from the outside stopcock via the mains pipe, which runs beneath the house and surfaces near the kitchen sink. The kitchen is the first port of call because the strongest demand for water is here, and the kitchen sink is traditionally where fresh drinking water is supplied. The mains pipe then rises to feed the boiler for the hot-water supply and all the cold-water taps in the bathroom. The toilet is also fed by mains pressure. The boiler feeds the unvented cylinder, and the hot water (under pressure) feeds all the hot-water taps in the house. Any secondary bathrooms will be supplied in the same way.

👍 **TOP TIP If you're planning to fit a boiler and cylinder, check that the sizes you are fitting are adequate for meeting the hot water requirements of your household, including any future expansion in demand, such as loft or kitchen extensions, or maybe an en-suite bathroom.**

DRAINAGE AND WASTE SYSTEMS

Three types of plastic can be used for external drainage and waste pipework. Acrylonitrile butadiene system (ABS) is a very tough plastic that can be used for both hot and cold waste. It can be connected using either solvent or compression joints. Polypropylene (PP) is a softer, more flexible plastic. It is impossible to glue PP, so the connections are always made using push-fit joints (see page 20). The most commonly used material for external waste pipes is unplasticized polyvinyl chloride (UPVC) – now that's what I call a mouthful! This type of plastic is damage-resistant to most kinds of household products like bleach and washing powder.

Although plastic pipes have been around for many years, until recently the most successful type has been waste pipe made from hard plastics. Plastic supply pipe (cold water) has now been introduced for use underground. Coloured blue, this medium-density polythene (MDPE) pipe is pressure- and corrosion-resistant, thank goodness, so you can fit it and then forget about it. Old mains pipes were quite often made of galvanized steel or lead, which eventually deteriorated. If you have a leaking old-style mains pipe on your property, make sure that you replace it with medium-density polythene. It is far more efficient than the old metal pipes and doesn't rot.

WASTE SYSTEMS

Once water enters the property, where does it all go every time you pull the plug on the washing-up, empty the bath or flush the toilet? Underneath every sink or bath there is a U-bend trap, which always retains a certain amount of water, preventing unpleasant sewer smells drifting back up the pipe – the water acts as an impenetrable barrier for such smells. The same principle is used for the toilet – each time it is flushed, everything is forced round the bend in the bottom of the loo, leaving enough clean water to act as a barrier that prevents smells.

SINGLE-STACK SYSTEM

1 vent
2 soil pipe
3 boss
4 reducer inlet

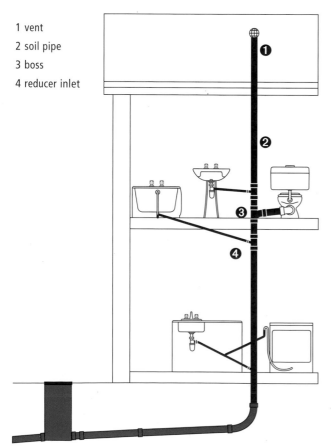

single stack

There are basically two types of drainage systems – a single stack or pipe, or the more common two-pipe system. In the former, all the soiled water and toilet waste enters a 100mm (4in) diameter soil stack pipe before running into the underground sewage pipe system via a manhole inspection chamber. This chamber allows for rodding to dislodge any blockages. The stack goes up to eaves level (guttering) to allow venting of the pipe. This single stack has to be well planned to prevent siphoning of any traps elsewhere in the system, which would allow smells or even sewer rats into the drainage system in your house.

two-pipe system

More common is the two-pipe drainage system, which is generally pre-1960s. This consists of a 100mm (4in) diameter pipe that takes the toilet waste directly via the manhole inspection chamber into the sewer. The pipe also extends to roof level to vent sewer gases and to prevent siphoning the water from the toilet trap. Bath and washbasin waste are often discharged into a smaller vertical pipe via an open hopper leading to a gully, into which the kitchen sink discharges independently. The gully branch enters the manhole inspection chamber, adjacent to the toilet waste branch, before heading off to join the main sewer. Again, the purpose of the inspection chamber is to allow access for rodding to clear any blockages.

DRAINAGE RESPONSIBILITY

Responsibility for maintaining drains up to the main sewer is usually the owner's. A block of terraced or semi-detached properties is often more complicated. If the properties were built pre-1937, the local authority is responsible for the cleansing, but if repairs are required, local authorities are empowered to reclaim the costs from the householder. Contact your local authority's technical services department to find out who is responsible for your drainage system.

Decades ago, installing a drainage system was a complex and skilled job that involved working with cement joints, salt-glazed clay pipes and fittings, and brick manhole inspection chambers. Today, most local authorities accept modern plastic pipes and inspection chambers, putting drainage well within the capabilities of the DIY enthusiast. Always check with your local authority before starting drain work, and have the work inspected and approved before reinstating the ground.

fitting and positioning pipes

Any pipe that discharges into a gully should extend down into the gully below the grille by about 50mm (2in). The reasons for this are twofold. If fallen leaves or rubbish cover the gully grille, water that is discharged from the house will be prevented from entering the underground drainage system. Also, when washing machines or baths discharge, there is a sudden exodus of water, and if that discharged water is not completely contained within the gully, serious problems may occur. Where water softens ground around and under the foundations of a building, for instance, undermining may result, which can cause major cracking.

So make sure that all waste pipes extend into the gully, and that there is a suitable mortar seal between the gully and surrounding surface.

TURNING OFF THE WATER

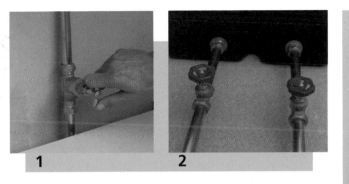

1   **2**

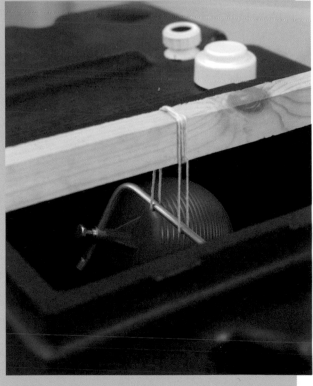

TOP TIP To temporarily restrict the tank in the loft from refilling, lay a wooden batten across the top of the tank, and tie the ball valve float to it. With the valve closed, open all the taps to empty the tank. Remember, this is temporary, until you fit a gate valve.

This may sound like stating the obvious, but turn off the water before you attempt to do any work in the bathroom. Locate the relevant valves and turn them off – there is normally a main stopcock under or near the kitchen sink **1**. On older, unimproved properties, this may be the only valve for the whole plumbing system; if so, you will have to turn this off and drain down the system by opening all the taps and flushing the toilet until the storage tank is empty.

GATE VALVES

You can avoid the need to empty the storage tank each time you work on the water system, and also keep the cold mains on, if you fit gate valves to both the cold feed pipes leaving the tank **2**. This straightforward job will enable you to leave the cold supply to your kitchen for fresh drinking water and cooking, even while you isolate the bathroom for carrying out repairs and alterations. And you won't need to wait for the empty storage tank to refill afterwards.

making plumbing connections

Connecting pipework is a fairly simple but rewarding task. Using solder and a gas torch may look complicated, but in reality the job is very straightforward. With a little practice and some basic common sense about safety when working with a naked flame, you can soon learn how to make decent solder joints. Using compression joints is like child's play in comparison, but the fittings do cost more to buy.

CONNECTING PIPEWORK

Joining two lengths of copper pipe, or fitting an elbow to turn a corner, is very simple for the amateur to do using a compression fitting. As the name implies, this type of fitting uses pressure to make a watertight joint, by tightening a cap nut and body to squeeze a round-edged brass ring (known as an olive) onto each pipe end.

the water supply must be in the off position if you cut or join any pipe

USING COMPRESSION JOINTS
The only tools you really need are a measuring tape, a tube cutter, two adjustable spanners, wire wool, and a roll of PTFE tape. Copper pipe can be cut with a hacksaw, but

unless this is done very carefully it will leave burred edges, making it difficult to get the olive onto the pipe.

Preferably, a tube-cutting tool should be used to cut the pipe **1**, because this will give you the perfect clean-cut and slightly tapered edge that ensures a good joint. The light taper enables the joint fitting, whether it is a compression or soldered fitting, to be pushed on snugly. On the side of the tube cutter there is a pointed reamer which is inserted into the end of the cut pipe; when twisted, it cleans off any sharp metal burrs inside the cut end of the pipe **2**.

Using wire wool, clean up the two cut ends of the pipework and the olives, and temporarily attach the two joint fittings. This will allow you to take an exact measurement of the pipes.

Slip the cap nut over the pipe and pop on the olive. The bevels on an olive are normally equal – if they are

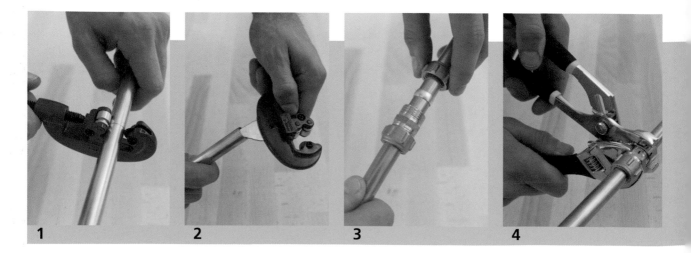

1   2   3   4

6

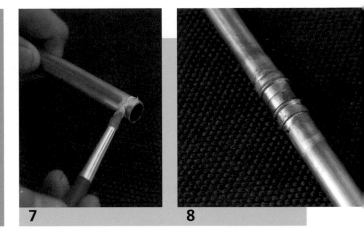

7   **8**

different, the longer bevel should point away from the cap nut **3**. Twist the pipe onto the joint body to ensure it's tight up to the body stop. Tighten the cap nut as tight as you can – avoid crushing the olive by over-tightening, which would cause the joint to leak. When hand-tight, use a marker pen to mark the cap nut and body in line. Hold the body steady with an adjustable spanner, and use the other spanner to give the nut a complete 360-degree turn **4** until the marks are parallel again **5**. This should make the joint perfectly watertight.

USING A SOLDERED FITTING

A tube-cutting tool should again be used to cut the pipe because it provides the ideal clean-cut and slightly tapered edge for the job. When using soldered fittings, clean up

the pipe ends and the inside of the fittings thoroughly with wire wool; to create a perfectly soldered joint, it is essential for copper to be absolutely clean.

For a soldered joint you have to use a paste called flux, which is a chemical cleaner that creates a barrier against oxidization until the solder sets. I recommend using capillary joints with integral rings of solder, which, when heated with the gas torch, release the correct amount of solder to seal the joint watertight. These joints are known in the trade as 'Yorkshire fittings'. Once you have applied the flux to both surfaces that are to be soldered **7**, apply the gas torch and heat all around the joint.

When the solder appears at the edge of the fitting in a bright silver ring, turn off the gas torch. The joint fitting is now sealed and watertight **8**.

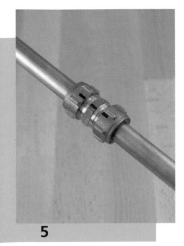

5

never leave a burning gas torch on – always turn it off after each joint is completed

TOMMY'S ADVICE: SAFETY

When using a gas torch, be aware of the dangers. Ventilate the room by opening the window and, when working in floor, wall or ceiling voids, place a fireproof barrier behind the pipe area you are soldering. Plumber's merchants supply fire-resistant fibreglass mats for this purpose, or you could use an old ceramic tile or a piece of asbestolux fireproof board.

plastic pipework for bathrooms

A versatile material, plastic is especially useful for carrying water. Plastic pipes are easy to use, lightweight and maintenance-free. Some types designed to withstand high temperatures are now available for hot water and heating systems, but most domestic plastic pipework is used for taking away waste water, especially in bathrooms.

TYPES OF PLASTIC PIPEWORK

Three types of plastic pipes are used for the water supply and heating installation in your house. Chlorinated polyvinyl chloride (CPVC), polybutylene (PB), and cross-linked polyethylene (PEX), pipes are tough and versatile enough for use as hot- and cold-water carriers, and also for central-heating systems.

The majority of bathroom waste pipes are made from polypropylene (PP) or acrolonitrate butadiene styrene (ABS for short). Waste pipes are 32mm (1¼in), 40mm (1½in) or 50mm (2in). They are joined to traps and other components using compression fittings, which contain a flexible sealing ring that makes dismantling for the removal of any blockages an easy task. Correctly fitted, a plastic compression joint should only need to be hand-tightened to create a watertight seal. Polypropylene (PP) pipes are joined together using push-fit connectors, while ABS pipes are solvent-welded.

Check the manufacturer's instructions with regard to expansion, which occurs in 100mm (4in) diameter soil pipe and may occur in plastic waste pipe carrying hot water over a distance. An expansion coupling, with a push-fit seal at one end to allow movement, may be necessary.

FITTING A PUSH-FIT JOINT

Cut the pipe to length using a junior hacksaw **1** or plastic

pipe cutters **2**. Using a file, chamfer down the outer edge of the pipe **3**. Remove any internal burrs using a craft knife **4** then apply a little silicone lubricant **5**. Push the pipe into the socket of the fitting, right up to the stop. Using a marker pen, mark the pipe, then withdraw the pipe 9mm (⅜in), using the mark you made as a measuring guide **6**. This is to allow for expansion caused by hot water.

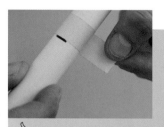

👍 **TOP TIP To help make a straight cut, wrap some masking tape around the plastic pipe to align with the measured mark.**

FITTING A SOLVENT-WELD JOINT

Some pipes are connected using fittings which are welded together with a purpose-made solvent. After fitting the waste run together without the solvent, simply reassemble,

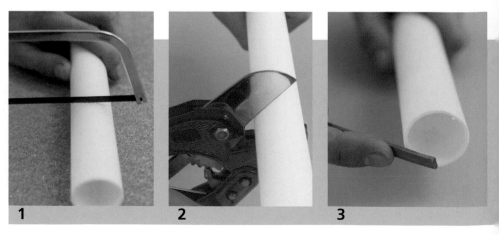

1   **2**   **3**

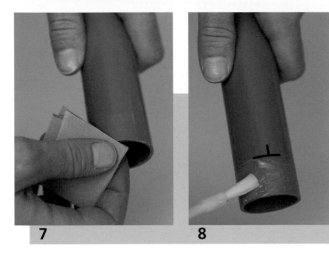

7   8

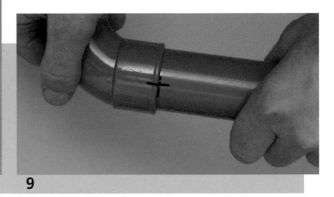

9

applying solvent to the inside of the socket and the outside end of the pipe. As for a push-fit joint, push the cut pipe into the socket to check on the fit, and mark the pipe at the end of the joint with a marker pen. For extra adhesion, use fine sandpaper to key the end of the pipe and the inside of the socket **7**. To assist the final positioning when the solvent is applied, mark an alignment line on both socket and pipe. When the solvent has been applied **8**, immediately align your marks and push the pipe into the socket with a twist **9**. You have to work quickly, as the joint will be cured in about 15–20 seconds. It will be ready for use with cold water in about one hour and with hot water in approximately four hours.

PEX PIPEWORK

I believe the future of plumbing is going to be based on PEX pipes. This plastic pipe is ideal for hot and cold water supplies as well as central-heating systems, particularly

👍 **TOP TIP If you follow the instructions, you should have a leak-proof waste run. However, if a joint is leaking slightly a repair can be effected by simply applying a touch more solvent to the mouth of the socket to seal the joint by capillary action.**

underfloor heating systems. The PEX system is simple to use and install. Already popular and successful across the USA and Europe, it is beginning to take off here as well, because it is so reliable yet easy to work with. Without any rubber seals or moving parts, it becomes a uniquely innovative design. The technique for jointing – by means of purpose-designed plastic socket-type fittings – is straightforward and safe, and without the fire risk associated with using a gas torch for soldering. A series of purpose-made adapters make the PEX system compatible with any existing conventionally piped system.

4   5   6

electricity & your bathroom

From a safety point of view, it is common sense to switch off all the power before commencing any work. Nowhere is this advice more valid than in the bathroom, where the combination of water and electricity CAN lead to tragic results.

ELECTRICITY RULES

Before undertaking any electrical alterations in your home, it's essential to understand how it all works. If you're well acquainted with the consumer unit and the electrical installation of your home, you may be content before starting work to simply turn off the mains switch, remove the relevant fuse or fuses for your bathroom and restore power to the rest of your home by switching the power back on. However, from a 'safety first' point of view, I recommend leaving the power supply in the 'off' position while performing **any** repairs or alterations.

Electricity supplies reach the house through a company mains cable, which enters the house by means of an underground or overhead armoured power cable. This supplies the current through the meter (which records how much you use) onto the consumer unit **1**. The consumer unit is the box containing all the fuse ways that protect the individual circuits in the house. The main on-off switch is located here, enabling you to isolate the power supply to the whole house. When fitting a new consumer unit, it's advisable to fit one a little larger than required, with some spare (unused) fuse ways for any additional circuits that may be required in the future.

The fuse ratings at the consumer unit for the various power requirements in the home are as follows:
⊙ Lighting circuits – 5 or 6 amp
⊙ Ring main circuits – 30 or 32 amp
⊙ Shower, cooker circuit – 32 , 40 or 45 amp

LIGHTING CIRCUIT

From the consumer unit the cable is fed to a room or series of rooms, normally through the ceiling voids. The circuit cable runs from one lighting point to the next, terminating at the most remote fitting. A switch cable is wired into each lighting point at a ceiling rose or junction box.

For lighting, the switch must be fitted on the flow side of the circuit rather than the return, in order to isolate the power from the light completely. If a switch were fitted to the return cable, it would still break the flow and turn out the light when switched off, but the light fitting would remain live and therefore potentially dangerous.

RING MAINS

The electricity supply to the sockets or appliances is called a ring main, with both ends of the ring cable connected to the live terminal of the circuit fuse/MCB. The lighting circuit is a radial circuit, which ends at the last lighting point. Each floor of a house usually has a separate ring main and a separate lighting circuit, allowing sections of the power supply to the house to be isolated.

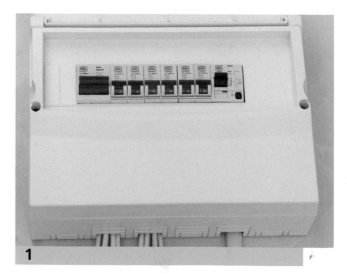

1

TOMMY'S ADVICE: SAFETY

✖ We all know how dangerous electricity can be. Water is an excellent conductor of electricity – combine the two incorrectly, and you could create a potentially lethal cocktail, so take care at all times.

✖ Stringent safety guidelines must be followed at all times regarding anything electrical to be used in your bathroom.

HERE ARE A FEW SIMPLE RULES TO FOLLOW:

✖ The only socket permitted in a bathroom is a special shower socket that must conform to BSEN 60742 chapter two, section one.

✖ No electrical appliances, such as hairdryers, portable fires etc, should be used in the bathroom, even if they're plugged in outside.

✖ Light switches must be ceiling-mounted with pull cords. Conventional switches must be fitted outside the bathroom. The same is true for switches controlling a shower or wall heater.

✖ Light fittings must be self-contained, shielded and mounted close to the ceiling, rather than a pendant fitting. You must not be able to reach out and touch any fitting from the bath or shower.

✖ Any bathroom heater must comply with IEE wiring regulations and be positioned in a safe place.

✖ A shower in a bedroom it must be no nearer than 3m (10ft) to the nearest socket outlet, which must be protected by a 30 milliamp RCD.

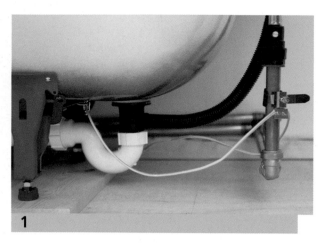

1

Being able to isolate the power to the bathroom while alterations are carried out is very handy. The power is normally delivered through $2.5mm^2$ twin-and-earth cable, with the live conductor covered in red protective plastic insulation and the neutral in black, with a bare earth conductor in between. All three conductors are encased in a thick white PVC protective outer sheath. Beware, it's not nail or screw proof! To avoid puncturing the cable, mark out all the cable runs on your wall before fitting your bathroom.

EARTH BONDING

All bathrooms require supplementary bonding, which is known as cross-bonding. For domestic use, this is a single $6mm^2$ copper wire cable, sleeved in green-and-yellow PVC **1**, which must be attached to non-electrical metal components in order to bond them to earth. The reason for this supplementary bonding is simple – safety! In a bathroom there are many non-electrical metal parts, such as baths, basins, showers, pipework, windows etc, any of which could cause a fatal accident if there were an electrical fault and they became accidentally live. Cross-bonding earths them safely if this happens.

ZONING

The Institution of Electrical Engineers (IEE) regulations clarify safety precautions when altering or rewiring a bathroom. The zoning recommendations are a very useful and simple guide stating that a room containing a bath

all bathrooms require supplementary bonding, the reason is simple – safety!

tub, and/or shower must be divided into zones **2**. They also give guidance on what should go where, including the position and direction of cable runs. Secondary bonding is not required outside these zones. Any light fittings selected must be suitable for the zone it's going to be fitted in. Follow the manufacturer's instructions closely, and measure the zones carefully.

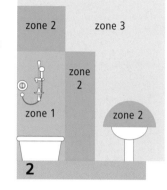

Any appliance rated above 4 is acceptable, as it indicates that it has been made with a higher water-resistance specification. Any electrical fitting to be installed in zones 1 and 2 must be made with the correct protection against water splashing, and coded at least 1Px4. Light fittings in zone 2 must be splash proof and have a rating no lower than 1Px4 (1 plus two digits denotes dust/moisture resistance indicating environment suitability). Light fittings in zone 1 also require protection from a 30mA RCD. A ceiling-mounted pull-switch can't be in either zone 1 or 2; it has to be either in or beyond zone 3.

CEILING HEIGHTS

Ceiling heights are also very important from an electrical-safety aspect. If the bathroom ceiling is 3m (10ft) or higher a ceiling-mounted pull switch may be positioned anywhere. If the ceiling height is 2.25–3m (7–10ft), the pull switch must be positioned a minimum of 610mm (48in) away from the bath or the shower on a horizontal plane. If the ceiling is lower than 2.25m (7ft), the switch must be positioned outside the room.

👍 **TOP TIP For any large electrical alteration or installation, get a qualified electrician to check out anything you're not sure of and to make sure that your work is safe.**

bathroom repairs

To save yourself a major headache later, carry out repairs as soon as the problem occurs. This will help to reduce the damage, because water leaking causes huge damage very quickly, particularly to timber, and could turn into a major renovation job!

leaks in the pipework

Plumbing leaks can be a real nightmare – they're extremely damaging to property, which can be very costly to repair. Fortunately, most of us only ever encounter problems with the pipework during extreme weather conditions, or when humans have interfered with it!

LEAKING PIPEWORK

Basically, four types of material have been traditionally used over the years for plumbing work. Lead was used extensively during Victorian times but was phased out after the Second World War, after the potential for lead poisoning was recognized. No longer used for new pipework, only repairs to existing lead plumbing are carried out nowadays.

Galvanized steel took over from lead and was popular for a time, but it rusts from the inside, which can block the supply. Another drawback is that when copper is connected to galvanized steel, a bad electro-chemical reaction takes place. Today, by far the most common form of supply pipe is copper, but this is now being gradually superseded by plastic.

👍 **TOP TIP Some old houses still have lead pipework. Water standing in lead pipes may absorb some toxic minerals, so if you haven't replaced your lead pipework or rising main yet, always let cold water run for a while before you draw it off for drinking or cooking with.**

DETECTING A LEAK

Not all leaks are caused by someone hammering a nail through the water main, although this is very common. The other usual type of damaging leak occurs in the winter, when a sudden drop in temperature causes the water in pipes to freeze. As the water freezes it expands, which fractures the pipe. When the temperature rises again the frozen water thaws and pours through the split.

Frozen pipe problems normally occur where pipework is exposed to freezing conditions such as in the loft. Leaks here cause so much damage because the water runs down through every ceiling on its way to the ground.

The worst damage occurs when owners are away, as the burst pipe may be left running for days before it is discovered. It's very important, therefore, to insulate all pipes and tanks in cold spaces efficiently. For a relatively small sum, you could save yourself thousands of pounds and an awful of a lot of grief.

Not all leaks manifest themselves so obviously – there could just be the sound of dripping water, or a damp patch, or, more ominously, a swelling or bulge slowly developing in the ceiling. If the latter happens to your ceiling, quickly gather lots of towels, buckets and pots, and use a screwdriver or something similar to make holes in the ceiling through which the water can drain into the buckets. This quick action could save your ceiling, and you can fill the holes with filler when the leak has been repaired and the plaster dried out.

If you detect a leak, turn off the water and drain the system in order to effect a repair (see page 17).

REPAIRING COPPER AND LEAD PIPES

If you discover a frozen pipe before it has fractured, you may be able to thaw it out with a hairdryer **1**, working it carefully along the affected area. Alternatively, use a hot-water bottle wrapped around the affected pipe **2**. Make sure that the tap from the frozen pipe is left open during the defrosting process. If the pipe has burst, a repair will need to be made. Although lead expands and contracts more readily than copper, you may need to temporarily repair a small split or pinhole in a lead pipe. Try closing it by burring the lead over. A burr is the rough, raised edge that is left on a piece of pipe when it has been cut, sheared or broken in any way. Because lead is such a soft

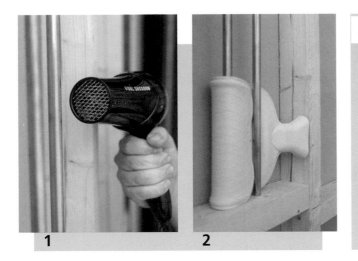

1 **2**

material you should be able to gently manipulate it, using a hammer or other blunt instrument, to cover the leak and stop it for a while. However, this is a very temporary measure, and you must replace the section of lead with copper or plastic pipework as soon as possible.

If you need to make a temporary repair on a damaged section of copper pipe, you can cut a short length from your garden hose. Using a craft knife, run a split from end to end and slip it over the damaged pipe **3**. Secure it in place using stiff wire or a few Jubilee clips. If the pipe has split but is still frozen, fit the hose before defrosting to avoid having to drain down the whole system.

There are a couple of proprietary products designed for temporary repairs: one is an amalgamating tape which

leave the tap from the frozen pipe open during defrosting

you bind around the damaged pipe **4**; and the other is a metal clamp which you simply screw onto the pipe **5**. However, I find it quicker and easier to simply replace damaged pipework or effect one of the above repairs.

If a fitting has been forced off the pipe during a freeze, the system will have to be drained and the fitting replaced. If it is a compression fitting, attempt to tighten up the existing one first before replacing it with a new one (see page 18).

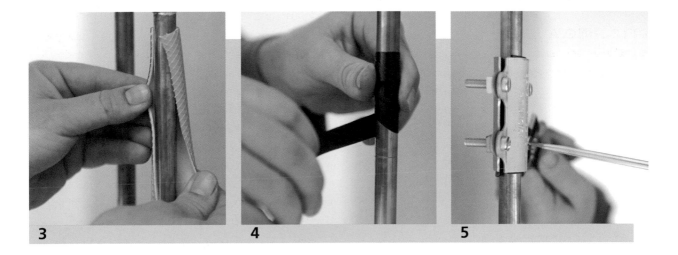

3 **4** **5**

repairing dripping taps

It can cost a small fortune to repair the damage caused by a pipe leak that is left unchecked, and everyone agrees it makes good sense to address such a problem as soon as it is noticed. With a dripping tap, however, many householders are content to let the problem go on for months or even years, even though it can cause unsightly staining to the tap itself or the appliance which it supplies, and is a waste of water. Yet fixing a faulty tap is much easier than most people think and a whole lot cheaper than calling out a plumber.

DRIPPING TAPS

It's amazing how many people seem to ignore a dripping tap. I think the reason must be that a tap may seem like a complicated piece of equipment to take apart, but in fact it is relatively straightforward. On a normal tap the leak is probably caused by one of three things:

◆ Dripping from the spout indicates a washer problem.

◆ Leaking from the head when the tap is running suggests that the O-ring or gland packing requires replacing.

◆ An old tap may be worn out at the seat and will continue to drip after the washer has been replaced.

REPLACING A WASHER

The only tools you're likely to need for this job are a slotted screwdriver and an adjustable spanner. Remember to turn off the water supply at the stopcock (see page 17).

Fully open the tap to drain any water before you begin to dismantle the tap. Prise off the tap head cover **1**. The shrouded cover needs to be unscrewed **2** to expose the headgear nut that is directly above the body of the tap **3**. Undo the nut with a spanner **4** then lift out the complete headgear assembly.

The washer is fixed to the jumper, which fits nicely in the bottom of the headgear. Depending on the type of tap, the jumper is either removed together with the headgear, or sits inside the tap body. Prise the washer from the jumper using a screwdriver **5** or undo the retaining nut to release it for replacement. Use penetrating oil to ease the nut if it's stuck; if the nut won't budge at all, you will have to replace both the jumper and washer. Replace the washer and reassemble the tap.

always leave the plug in so bits don't disappear down the hole

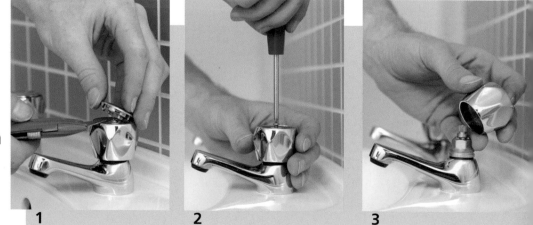

1   2   3

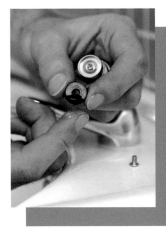

👍 **TOP TIP** It's always worth keeping a few spare washers handy in your fixing kit. To ease a stiff washer on or off the tap's innards, lubricate it with silicone grease. To soften a new washer and make it easier to fit, soak it thoroughly in hot water.

REGRINDING THE SEAT

If you are unlucky and the tap still drips after you have changed the washer on the tap correctly, the tap seat is probably worn and so letting water through. To rectify this a special reseating tool can be acquired from a plumber's merchant to regrind the seat flat. Simply remove the headgear and jumper again, then screw the reseating tool into the tap body. Bring the cutter into contact with the tap seat and turn the handle to re-cut the worn seat surface smooth and flat **6**.

If that all sounds too difficult (and it really isn't!) you can purchase a nylon liner, which is sold with a matching jumper and washer **7**. Just pop the liner over the old tap seat, fit the replacement jumper, reassemble the tap and then close it. This action forces the liner into position and prevents the tap from leaking.

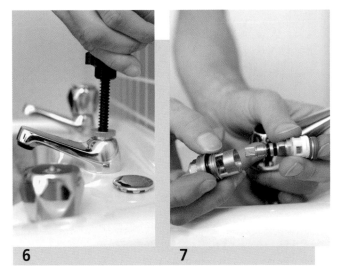

6　　　　　**7**

REPLACING A CERAMIC DISK CARTRIDGE

These are much simpler to repair than conventional taps. Turn off the water supply as before. Unscrew the old cartridge from the tap and fit a replacement. Note that cartridges are left- or right-handed on mixer taps.

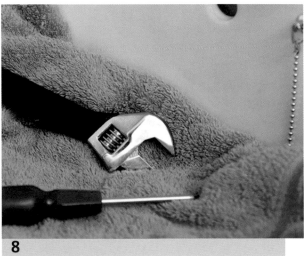

8

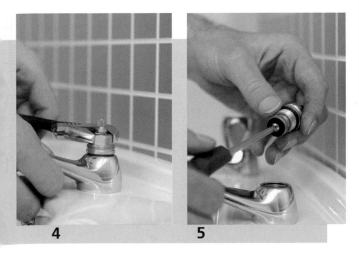

4　　　　　**5**

👍 **TOP TIP** Whenever you are servicing a tap, leave the plug in to prevent any vital components disappearing down the hole. It may also be wise to put a folded towel in the sink while you're working **8**, which will prevent any damage to the surface if you drop something weighty such as a spanner.

clearing blockages

Blocked waste pipes are not something that we like to spend too much time thinking about, and we all hope that they happen to someone else rather than to us. Unfortunately, this is not always the case. Even if you follow good practice at home, one of your neighbours might not be so careful, and the result may back up all the way to your own toilet.

SINKS AND BATHS

The biggest cause of blockages in waste pipes is hair, whether carelessly discarded after brushing or simply due to natural hair loss. A friend of mine was so concerned about his receding hairline (I have to admit the tide was going out extremely quickly in his case) that he would filter the bathwater as it drained to check how much hair he had lost.

Slow-draining water is the early warning sign of an impending blockage problem, so never ignore it. It's possible to clear a partial blockage at this stage with a proprietary chemical cleaner, which should be used occasionally anyway to keep the waste system free-flowing and problem-free. Be sure to follow the manufacturer's instructions, as these cleaners often contain ingredients that could be hazardous to the skin and eyes, and always wear protective gear, such as rubber gloves and goggles.

Alternatively, try a plunger. Simply block the overflow with a damp cloth then pump the plunger up and down over the plughole. For the plunger to work effectively, the rubber end must be submerged in water. Another useful tool is a suction pump, used in a similar way to the plunger. Place it over the plughole and push the handle down to produce a vacuum, pull it up sharply and the water should drain away **1**. If this approach fails to shift the blockage, you may have to undo the waste trap beneath the bath or sink to physically remove the blockage.

1

The waste trap assembly is usually attached to a horizontal waste pipe, which carries the waste water through the wall to the gully outside, via a hopper or stack waste system. There should be rodding eye plugs at various points along the pipe runs, particularly at angled junctions, which can be undone with a spanner to allow access. You could try to clear a blockage using some strong wire at one of these access points **2**.

2

👍 TOP TIP **Wire coat hangers, acquired from your local dry-cleaners, can be very handy to have around for various jobs, and this is one of them. Untwist the hanger to give you a long, stiff wire pipe-prodder, or bend the end to create a blockage puller.**

If you cannot clear the blockage using either of the above methods, you may have to rent a drain auger from a hire shop (see right). An auger may also be the only way to clear a blockage in the absence of rodding eyes in the waste-pipe line. Remove the waste trap and clear the blockage using the auger from the sink or bath waste-pipe end.

👍 TOP TIP **Prevention is always better than cure, so regular use of bleach and disinfectant will ensure effective, hygienic and problem-free waste systems.**

TOILETS

A toilet bowl that fills up with water before it drains away when you flush could be an advance warning of a problem looming, literally, just around the corner.

If the blockage is in the trap of the pan, try using a WC plunger to clear it. This is a larger version of the sink plunger and works in a similar manner, forcing the blockage along the waste and into the drains. Position the rubber end of the plunger right down in the neck of the U-bend and pump vigorously **3**. If the blockage clears, the water level will suddenly drop, usually accompanied by a loud noise.

If this fails to clear the blockage, you may have to use a toilet auger, which has a shaped head on the end of a flexible rod. Push this as far round the U-bend as possible **4**, then crank the handle to try to break up and clear the blockage. You can rent both a WC plunger and an auger from a hire shop, but remember to clean and disinfect them thoroughly before you return them.

Clearing a blockage in a cast iron soil stack may prove difficult, because the rodding eyes (see left) may be seized

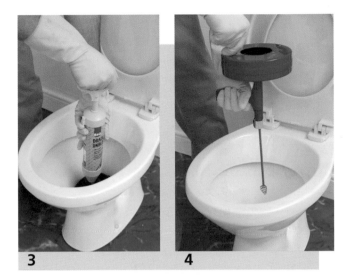

3   **4**

and rusted. In this case, the pipe may require cleaning either from the top, where the gases are vented over the roof, or from below. For the latter, you need to gain access via a manhole – a brick-built, concrete or plastic access chamber. Lift up the manhole cover, push the auger up the pipe until it meets the blockage then turn the auger handle until the blockage disperses.

A blockage in underground sewage pipes also needs to be accessed via a manhole. Most of these blockages can be cleared using drain rods (available on hire) with one of two fittings. A corkscrew head will clear most blocked drains, and a rubber plunger type is useful where the drain joins the main sewer. Attach the head to a couple of rods and push to-and-fro until the blockage clears.

TOMMY'S TIP

If someone has dropped an object in the toilet bowl, it has to be removed! Insert your hand into the bottom of a black plastic refuse bag so it's like a long glove, then reach into the water and pick up the offending article. As you lift it out, withdraw your arm from the bag, effectively folding the bag around the item for easy disposal – and no mess.

repairing toilets

Apart from the kitchen sink, the most used and overworked piece of plumbing kit in the whole house has to be the toilet, particularly if you have kids. I know this to be the case at home, just by the number of toilet rolls our family gets through each week – it's enough to keep a whole litter of labrador puppies amused for ages!

PROBLEMS WITH A TOILET FLUSH

Obviously, constant use of an appliance means increased wear, but regular maintenance checks should keep problems to a minimum. There are basically three types of toilet cistern: low-level (easy to maintain), high-level (accessible by stepladder) or concealed, which may present maintenance access problems if the installation wasn't well planned. The working components in a toilet's cistern are readily available for replacement or repair from a plumber's merchant or a DIY store.

👍 **TOP TIP When purchasing a replacement part, take the old one with you to make sure you are buying the right type. Also, write down the name and make of the suite. Even better, telephone ahead to avoid a wasted journey.**

HOW DOES A TOILET FLUSH?
The most common flushing system is operated by a direct-action cistern. Water enters the cistern via the supply pipe and is controlled by a valve. This valve is in turn controlled by a hollow plastic float on an extended arm, which opens or closes the valve as the water level inside the cistern changes. The water level is preset to the required volume necessary for flushing.

THE MOST COMMON PROBLEMS
A few problems can occur in a cistern, the most common being a faulty float valve or a poorly adjusted float arm. Evidence of either of these is clearly visible from the outside, when water drips or runs from the overflow pipe. When this occurs your cistern will be constantly filling, which is both noisy and wasteful.

float valve problems
Float valves are used in two types of situations – loft storage tanks and toilet cisterns. In a conventional system

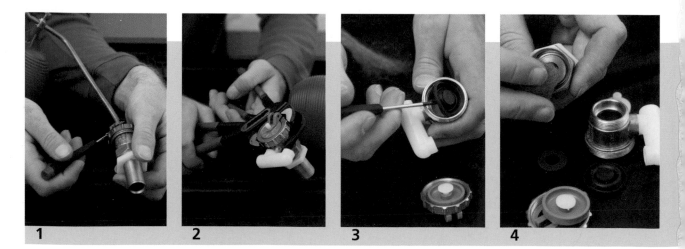

1   2   3   4

the cold-water storage tank in the loft supplies all the hot and cold water for the bathroom. If the overflow is running from this tank, you will need to shut off the water supply (see page 17) and change the washer in the float valve.

When you have shut off the water supply, turn on all the bathroom taps and flush the toilet repeatedly to empty the tank. Disconnect the valve from its supply pipe and remove it from the cistern. Remove the split pin from the valve **1**, releasing the float arm. Unscrew the cap at the end of the float valve **2**, remove the piston from the body and unscrew the piston end cap. To prevent the piston from rotating, insert a slotted screwdriver into the gap. Remove the old washer **3** and clean any debris off the cap using wire wool **4**. Fit the new washer, lubricating it with a touch of silicone grease **5**. Replace the piston, reconnect the float arm to the valve and replace the split pin.

An adjustment screw on the type of float valve used in the toilet cistern controls the water level within the cistern **6**. If, after reducing the level the valve still lets water get by, this is normally an indication that either the washer or the diaphragm, depending on valve type, needs replacing.

changing a washer

The most common float-valve system today uses a diaphragm valve, replacing the old washer with a large diaphragm that is less susceptible to wear and limescale. The newer valve appears to be much more durable and reduces maintenance.

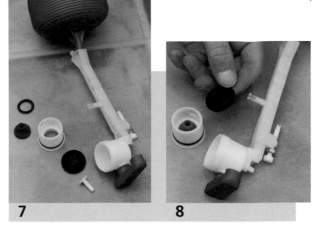

7   8

To change the diaphragm, isolate the water supply (see page 17) and drain the cistern by flushing the toilet. Undo the nut connecting the float arm at the top of the valve and put it to one side. Unscrew the ball-valve assembly, then remove the plastic piston **7**. Remove the worn diaphragm, clean off any residue build-up, and fit a new diaphragm **8**, adding a touch of silicone grease to act as a lubricant. Reassemble all the bits and turn the water back on.

replacing a flap valve

Having to press the lever repeatedly to get the toilet to flush indicates that the flap valve is probably faulty and requires replacing. To replace the flap valve, first tie the float arm to a wooden batten laid across the cistern, to prevent it refilling. Flush the cistern to clear the water and check the float arm is not allowing the cistern to refill. Unscrew the large nut that connects the flush pipe to the cistern, and move it aside.

Undo the siphon-retaining nut to the cistern base (some water will spill out at this point, so have some towels or cloths ready). Disconnect the flushing arm **9** and carefully ease out the siphon. Remove the diaphragm from the metal plate and replace it with a new one. Reassemble the flushing system, reconnecting the flush pipe to complete the job.

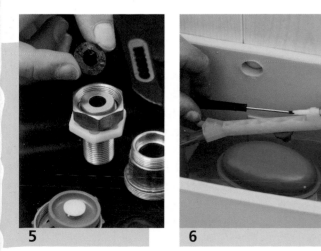

5   6

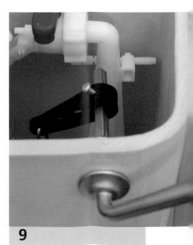

9

changing a radiator

If the need to replace a radiator should arise, the simplest thing to do is to find an exact copy. Alternatively, you could use the opportunity to install a heated towel rail. This should be fairly straightforward, although you will have to fix new hanging brackets and extend the pipework, which will involve draining down the central heating system rather than just the radiator.

REPLACING A RADIATOR WITH A TOWEL RAIL

Most modern towel rails are attached through the wall, rather than the floor, so you'll more than likely need to adapt the pipework. First, you'll need to drain down the system (see below), then disconnect and remove the old radiator complete with its valves **1**. Clean up the open pipes with steel wool and temporarily cap them off with a fitting to prevent any debris getting in while you work. Remove the old brackets. At this point it's a good idea to mark a centre line on the wall where the old radiator hung, to act as a guide.

ADAPTING TO A DIFFERENT SIZE
You'll need to gain access to the floor void to alter the pipework. Remove the skirting and lift the floorboards. For clarity the pipework is shown here in an empty void, but in reality you'll have to cut away some plasterboard above the cut pipe or even chase out a channel for the new pipe in a solid wall. Either way you will have to make good the wall before moving on to fix the towel rail.

Mark positions for the new copper tails (for both the feed and the return pipes) **2**, then cut out a section of noggin, or chase into a solid wall, to take the new pipework. Run the new pipework from the old pipes, using either compression or capillary fittings, and add two elbow joints so each pipe can go into the wall and exit higher up **3** ready to connect to the towel rail. Tape over the open ends to keep out any debris while you are restoring the wall, flooring and skirting.

INSTALLING THE NEW TOWEL RAIL
Unpack the towel rail and remove the plastic bungs that protect the threads. Leave on the protective outer coating until the towel rail is fixed in place – you are bound to get a few scratches as you install it. Unscrew the two valve adapters from the bottom of the old radiator with an adjustable spanner. Unscrew the bleed valve with a bleed key, then remove both blank plugs from the top of the new towel rail with a radiator spanner **4**. Screw the adapters and plugs into the new towel rail. Finally, screw the bleed valve into the blanking plug. The new towel rail is now ready for fixing to the wall.

The towel rail is is attached to the wall in a different way to an ordinary radiator. The bottom should be level with the skirting board. Using a spirit level, draw a horizontal guide on the wall for the top brackets **5**, and fix them in place **6**. Repeat for the bottom brackets. Now you can connect the valves to the supply and return pipes. Check the position of the towel rail on the brackets.

DRAINING AND REFILLING THE SYSTEM
Shut off the boiler, switch off the fuel supply and leave to cool down for a few hours. Once the water in the heating system is cold, cut off the water supply at the expansion tank, either by closing the tank's gate valve or tying up the float arm. Fit one end of a garden hose over the drain cock, usually on the return pipe at the boiler, with the other end over an external drain gully. Open the drain cock using an Allen key or spanner, and let the system drain down completely. Pack towels or cloths around the valve in case of spillages. To refill the system, close the drain cock and restore the water supply.

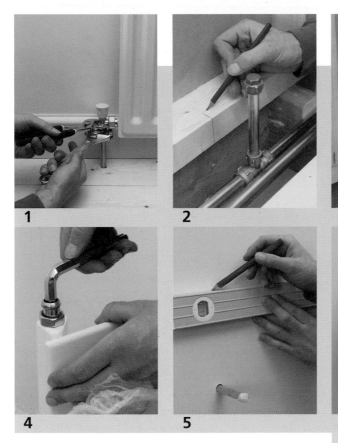

1

2

3

4

5

6

7

Hold the valve to the pipe, mark the pipe and cut to fit **7**. Connect the towel rail and cover the fixings with the clip-on caps. Connect the valves to the adapters and close the drain cock. Restore the water supply to the expansion tank and refill the central heating system (see opposite). Fill the radiator by opening the valves, bleeding any air in the radiator through the bleed valve **8**.

👍 **TOP TIP Hold a cloth to the bleed valve as you are screwing in, to catch any water that might leak out 9.**

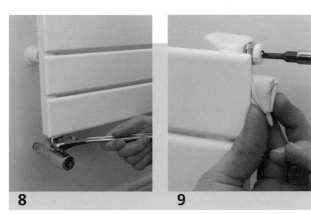

8

9

renovating your bathroom

If you wish to renew your bathroom, but you simply can't afford to rip everything out and start from scratch, it is still possible to renovate your bathroom on a relatively small budget yet create a wonderful transformation.

A COMPLETE OVERHAUL

A tired bathroom can often be completely rejuvenated without having to replace any of the sanitary ware. Clean, fresh paint and bright new fittings will give it a new lease of life for a fraction of the cost.

First, clean the tiles thoroughly, and I mean thoroughly, then renew the tile grouting with either new grout or a renovator. Change the old taps **1**, waste and overflow for new chrome ones or install a mixer. Alternatively, take the old fittings, if they are good enough, to your local plumber's merchant and have them sent away for overhauling and re-chroming. With the taps and fittings off, thoroughly clean all the surfaces of the toilet, bath and washbasin, removing any limescale staining. Take this opportunity to renew all the silicone mastic seams, then replace the taps and fittings **2**.

Install a new light fitting and repaint the ceiling, walls and woodwork. To add colour to a plain white tiled wall, paint a strip with one of the new products available, or use transfers or even stick-on tiles (see page 39). Replace the floor covering with an inexpensive piece of vinyl, and perhaps fit a new venetian-style blind. Change the toilet seat. There you have it – a new-look bathroom **3**.

CLEANING AND REPLACING TILES

A tiled wall should last a very long time without needing to be replaced, so you can imagine my concern and extreme annoyance when cracked and damaged tiles suddenly started to appear in the bathroom at home, on a wall that I had fairly recently tiled. I quizzed my wife and three children one by one, but needless to say, they all denied any knowledge or responsibility, and the mystery remains unsolved to this day. These things do happen, and one might be excused for assuming that the seismic effects of the great shifting plates of California had reached Hackney. Of course, I was left with the job of replacing the damaged tiles. Fortunately, repairing a tiled wall is fairly straightforward and restoring a tired one is even easier, although it does demand some physical effort.

Apart from incidental damage, tiles are only usually replaced for a more fashionable style, or because they're very old. When tiles age the glaze tends to craze slightly and they begin to look ugly as they are covered in black, circular or angular lines. When this occurs, it's probably worth considering taking the lot off and starting afresh.

If the tiles themselves are sound and still of an acceptable style, but the grouting is black and unsightly, it may well be worth renovating them, because all that's required is a very small outlay and a bit of elbow grease.

CLEANING A TILED WALL

Mix a bucket of bleach and warm water in equal proportions – this is a very strong solution, so wear protective rubber gloves and goggles. Apply it to the tiles using a two-sided sponge, with a soft surface on one side and a pan scrubber on the other. Using the rough side, scrub the tiles and grouting joints to remove any condensation mould marks, which are normally black **4**. Polish the tiles dry with a soft cloth **5**.

👍 **TOP TIP To help avoid the problem of condensation, which causes the build up of unsightly sooty mould in bathrooms, shower or bath with a small window open. If that's too cold, open the window immediately after you have left the bathroom, and leave the bathroom door open to allow fresh air to circulate.**

REGROUTING A TILED WALL

Where grouting has become discoloured, scrape it out a few millimetres (⅛in) below the surface and apply fresh grout. You can use a special tool **6**, or you could make one by reshaping the end of an old screwdriver with a grinder. Be careful not to slip and scratch the face of a tile. The grouting might be very tough, and it may be necessary to tap the tool along the joint with a mallet or piece of wood, such as a length of 50 x 50mm (2 x 2in) batten.

Once you've brushed out the dust and debris from the joints, wash down the tile surface with sugar soap then polish the tiles dry with a soft cloth. Alternatively, if the tiles were washed clean before you removed the grouting, use your vacuum cleaner to suck up all the dust before you regrout. Grout all over (see pages 39 and 102).

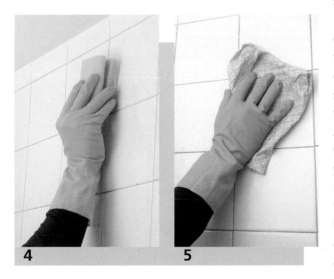

4

5

6

REJUVENATING GROUT

Instead of raking out the old grouting, you can use a grouting renovator. This comes in the form of a ready-mixed paint which forms a waterproof bond over the old grouting, but does not stick to the tiles. If this is your preferred option, first wash the tiles and grouting down with sugar soap, and allow the tiles to dry fully. Paint the renovator over the joints **1** and leave to dry, which should take a couple of hours. Spray the whole surface with water, wait three minutes, then wipe off the excess with a damp sponge. Dry and polish the tile surface with a soft cloth to finish, and the job is done.

REMOVING DAMAGED TILES

If one or two tiles on a good wall have been cracked or damaged, do not despair. Providing you kept the half-box you over-ordered at the back of the shed, along with all the half-empty cans of paint, this is a relatively easy problem to overcome.

The first thing to do is to remove the grouting from around the tile, or tiles, you intend to replace (see page 37). Using an electric drill fitted with a masonry bit, drill a sequence of holes near the centre of the tile **2**; this will weaken the tile considerably. Make sure that you only drill through the tile, and try not to damage the wall behind. Using a hammer and small, sharp cold chisel, and working from the weakened centre to the outside edges, carefully break out the tile **3**. It's worth taking the time to sharpen the cold chisel on a grinder before you start.

👍 **TOP SAFETY TIP Flying ceramic slivers could very easily blind or cut you, so always wear goggles and gloves for this task. Take your time over the removal process, to avoid damaging any of the surrounding tiles.**

Next, remove any old adhesive and brush off any dust, ready for fixing the replacement tile. Butter a bed of adhesive onto the back of the tile **4** and press it into position flush with the surrounding tiles **5**. Clean off any excess adhesive and leave to set before grouting.

6

but the best results will be achieved if you use a specialist tile paint, several types of which are readily available. You need to apply a special tile primer before you paint. It's always wise to research this option properly and test samples on a spare tile to see the finish and colour before you start the job. First, sand the surface to create a key for the paint. Clean off any dust with clean water, then dry thoroughly. Protect the surface around the area to be painted. Apply the primer and then the paint, either over the whole tiled surface or in a strip **7**.

GROUTING

You can use one of the ready-mixed grouts or buy powder and mix it with water to form a paste the consistency of double cream. Apply the paste to the tile surface, using a tiler's rubber float to press the grout into the joints **6**. Wipe the grout off with a damp sponge before it sets. When it has set, run something smooth, such as a plastic pen cover, over the joints to compress and shape them.

When the grout has dried, polish the tiles using a soft, dry cloth. Leave a newly tiled shower for a week before you use it. After doing such a good job, I think you deserve to treat yourself to a proper night out on the tiles!

PAINTING TILES

The most dramatic change you can make to unsightly old tiles is to paint them. You can use ordinary paint on tiles,

TILE TRANSFERS

If your tiles are in a good condition, but you want to liven them up, tile transfers are a simple solution. The tile surface must be absolutely free of grease, so clean them thoroughly as described on page 37. Soak the transfer in warm water for approximately 20 seconds, wet the surface of the tile you want to alter, then slide the transfer off the backing paper into the desired position **8**. Once it is correctly aligned, smooth out any wrinkles or bubbles with a dry cloth, then gently dab it dry **9**. If you want to use transfers in an area that is liable to get a lot of direct water, for instance in a shower area, apply a coat of clear, waterproof varnish over the transfers after application. Later, if you need to clean the transfers, use a soft cloth and a liquid detergent solution; never use an abrasive cleaner or bleach, as these will damage the transfer.

7

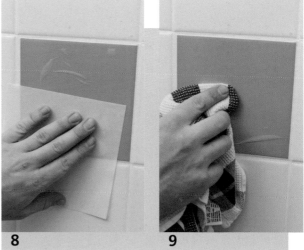

8   **9**

REMOVING AND REAPPLYING SEALANT

When grouting around a bath or shower, many people make the mistake of grouting right down to the edge of the bath or shower tray. Ideally, the last joint should be wiped out and allowed to dry, then a silicone bead applied along the length of the joint. This is because the tiled wall surface and the bath or shower tray, being made of two different materials, expand and contract at different rates, which would soon cause a static seal to fail. The silicone bead provides a flexible seal that stays intact and prevents water from seeping through and causing no end of problems. There are many different types of silicone mastic sealant for different applications. Always use a good quality one and make sure it is the correct type for the job.

From time to time a flexible seal may need to be replaced – a dark line between the silicone bead and the wall or surface indicates loss of adhesion. The sealant can often be removed by picking it off slowly in one long piece. Otherwise, use a craft knife to carefully cut along the tiled surface to break the seal and then peel it off **1**. Clean the two surfaces thoroughly in preparation for the new sealant, making sure that both surfaces are completely free from dirt and grease (see page 37). They must also be bone dry before you apply the sealant.

👍 **TOP TIP To ensure cleanliness and thus adhesion of the sealant, rub the surfaces to be sealed with a cloth soaked in methylated spirit. Make sure both surfaces are thoroughly dry and leave for an hour or so before applying the new silicone bead.**

Fit the sealant tube into the mastic gun then cut off the end of the nozzle at an angle to suit the size of the gap. Point the nozzle at one end of the joint, hold the gun steady and squeeze the trigger while moving the gun along. If there is a corner, work away from it **2**. Have handy a small bowl filled with equal parts water and washing-up liquid; dip your finger into the bowl and wipe over the bead to give a smooth finish along the length of the joint. Alternatively, dip the handle of a spoon or fork into the solution, then run it along the joint to shape it.

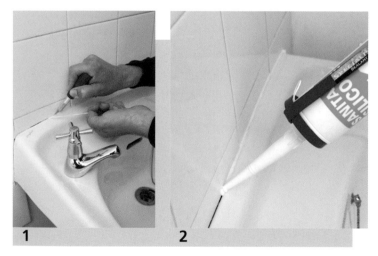

1

2

DESCALING

If you happen to live, as I do, in a 'hard water' area, the plumbing and heating systems can develop problems caused by the scaling up of pipework and fittings. The water companies supply us with clean, fresh water, with all the harmful impurities removed, but still containing minerals absorbed from the ground. It's the concentration of these minerals that determines how hard or soft our water is. The general rule of thumb is that flat areas, which

depend on supplies being drawn from underground, have a higher mineral content, therefore hard water. If you live in an area that draws its water from rocky surface terrain you are likely to have soft water.

The tell-tale signs of hard water are a scaled-up kettle or shower head, or staining on the bath or washbasin. This hard limescale also builds up on the inside of pipes. Hot-water cylinders are particularly vulnerable, with their efficiency being affected by up to 70%. Trying to prevent the build-up of limescale or staining, which can be very difficult to remove once established, is a constant but important battle.

Keep a check on limescale build up inside the shower head especially. An accumulation of limescale here will partially or wholly block the tiny holes, resulting in poor performance from your shower. To remove the scale, undo the shower head, disassemble the perforated spray plate **3** and soak all the components in a pot of proprietary descaler. Once the limescale has dissolved, rinse all the parts thoroughly with clean water, reassemble them, and

refit. Your shower should function perfectly once again. Descalent can also be used around the taps to help keep the bathroom gleaming.

If hard water is causing you real problems, consider installing a water softener to treat all your water, except drinking water. Fitting one can be difficult but is do-able depending on your proficiency.

3

BATH PANELS

A bath panel has a big impact on a bathroom, so installing a new one in a different style can have a dramatic effect, changing its appearance beyond all recognition.

REPLACING A BATH PANEL

Remove the old panel and use it as a template for your new panel. Lay it on a sheet of MDF set across a pair of carpenter's stools, or stand the sheet up if you are short of space, and draw around it to transfer the shape **4**. Alternatively, measure up the panel space you have just exposed and transfer these measurements to the MDF. With the MDF firmly secured, cut out the panel. Use a jigsaw and

cut a fraction outside the line **5**, then smooth off the excess using a hand plane to give an accurate fit.

👍 **TOP TIP Before cutting, double-check the dimensions of the panel, skirting and moulding.**

Next, fix on any matching skirting. Use PVA glue and pins, or screw from the inside to fix. Draw a line of equal margin all around the panel, cut and mitre moulding of your choice to match the lines, then glue and pin it in position **6**. With the panel still on the stools, paint it (including the edges) with a coat of primer, two undercoats and a topcoat. When dry, fix it into position. It'll make your bathroom look like new (see page 36).

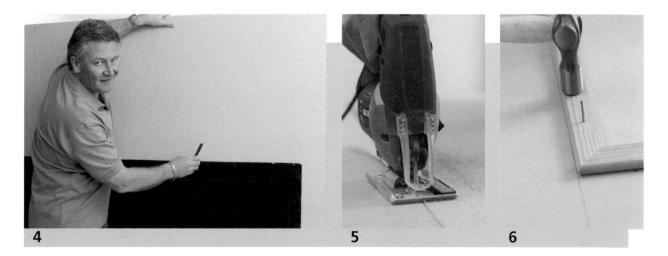

4 **5** **6**

dismantling a bathroom

Occasionally, you may have to replace a single item in your bathroom, but it is probably more likely that you will want to remove the whole suite to make way for a new one. Here are a few tips to help you strip out your bathroom.

STRIPPING OUT THE OLD

If you are replacing any item in your bathroom with a similar fitting, some of the pipework may need to be carefully dismantled for reconnection, but if you are rearranging the bathroom, you might as well just cut it all out with a hacksaw. Remember to turn off the water supply before you attempt to remove anything (see page 17)!

WASHBASINS

Basins come in a variety of designs. They may be supported on large, screw-fixed brackets, set into counter-tops, or balanced on a pedestal; while the pipework may be easily accessible, concealed within a wall, or hidden in a cupboard or behind the pedestal. If you want to reuse the existing pipework, undo the compression nuts on the tap connectors with a wrench or cranked spanner. Disconnect the trap waste, then undo and remove any fixings and wall brackets with a screwdriver.

👍 **TOP TIP If you're going to reuse connectors or old pipework, block the ends off with insulating tape or something similar, to prevent any debris dropping into the pipework and causing blockages later on.**

BATHS

Cut through the supply pipes and overflow with a hacksaw, and if the bath has adjustable feet, wind them down. Use a craft knife to cut through the mastic seal around the bath (see page 40). With many baths, you can just lean down on the bath and pull it away from the wall, but if there is a timber framework built around it to hold a panel, you'll have to dismantle this first.

Plastic and pressed-steel baths are relatively light, so they can be removed in one piece, but a cast-iron bath is a tremendous weight. Unless you wish to restore the bath, it may be a lot easier to break it into manageable pieces using a sledge hammer. This will be noisy, and dangerous because of flying debris, so wear ear protectors, safety glasses and gloves to complete the job safely. And don't forget to warn your neighbours before you start!

TOILETS

Flush the toilet, after you've turned off the water supply, to empty the cistern, then check it isn't refilling. Using a wrench, disconnect the supply and overflow pipes. Undo the fixings for the cistern. The pan is normally stuck or fixed with brass screws to the floor. Brass is a soft material, so be careful when removing them.

If the pan waste is connected to the soil pipe with a flexible push-fit pan connector, you can simply pull the pan free. If the pan is sealed with putty or, even worse, a sand-and-cement mix, this will have to be removed without damaging the cast-iron soil pipe collar, by carefully breaking out the pan at the outlet with a hammer. Block up the soil pipe to prevent any debris falling in and clogging the pipe; do this by stuffing in a plastic bag filled with old rags and tied to a length of string for easy removal. Break out the remaining pieces and jointing of the old pan using a hammer and sharp cold chisel.

If you're unfortunate and break some of the soil pipe collar, this can be rectified by cutting off the damaged section using a chain-link pipe cutter or a large angle grinder (both available on hire). Make a series of horizontal cuts followed by a vertical cut until you have a clean, straight spigot end. You can then connect the new pan using a push-fit flexible pan connector.

bathroom planning

What a choice! Any showroom these days can offer you an enormous range of bathroom styles, from taps to toilet seats. Take yourself down there and make that choice, after reading my advice first of course.

bathroom styles & layout

The most important part of creating a new bathroom is the planning – apart from paying for it, of course! Although the look is very important, a truly wonderful bathroom must also work on a practical level and be much more than merely functional.

THE PLEASURE OF A BEAUTIFUL BATHROOM

A well-planned, beautifully finished bathroom can make all the difference to your daily routine. It should always be a pleasure to use and cater for everything you and your family require, whilst avoiding a cramped feeling. Such a bathroom need not be hugely expensive, and it's a good idea to set yourself a budget and go from there, otherwise you might be tempted to overspend.

If space is a problem, and with most bathrooms it is, think about what you really use and what you could do without. Personally, I like having both a bath and a shower, but I could do without the bidet! The shower provides an essential energizing, head-clearing start to the day, while I enjoy a long leisurely soak in the bath when I have time to relax.

Ideally, and if we had the money and the space, we would all probably choose to have a separate shower room, or failing that perhaps a built-in enclosure within the bathroom. If there just isn't the room in your bathroom, and you don't want to lose the bath for the sake of a shower, you may have to compromise by showering in the bath. This can often be achieved quite simply by changing the bath taps for a shower mixer unit, with an extended hose mounted on the wall.

Once you have sorted out all the practicalities, you can decide on style and colour. Personally, I think bathroom suites shouldn't be any colour other than white. They can be off-white, or even ivory, but they must be white – ageless and easy to clean.

BATHS

When it comes to choosing a bath, you don't have to stick with the one that comes with a matching toilet and basin. You can select an individual bath to fit in with the other bathroom fittings and the overall decor. There are hundreds of different shapes and styles to suit every taste, as well as a wide choice of materials, including acrylic, resin, steel and cast iron. I have even seen a mind-boggling glass bath for sale!

3

1

2

4

Over the years, though, I have found that tried and tested materials always work best, and for me nothing can beat a cast-iron bath for heat retention **1**. Cast-iron baths can be very heavy, though, depending on their quality, so if you have set your heart on one of these, you need to make sure that the floor joists are strong enough to take the extra weight. Acrylic baths are lightweight by comparison and come in a vast range of profiles **2**. Corner ones are often the practical answer where space is at a premium. The thicker the acrylic, the more sturdy and durable the bath itself will be.

If you are really into self-pampering, then a spa bath or jacuzzi may be the answer **3**. Essentially, these baths incorporate an air-jet system that provides massaging options. Some even have underwater lighting **4**.

A major decision is whether the bath is to be fitted with panelling to suit the decor, or freestanding. With the former all the pipework will be hidden **5**, but with the latter the pipework can become a feature in itself **6**.

5

6

WASHBASINS

There are so many washbasin designs on offer, with a comparable choice of prices, that you will be completely spoilt for choice. Most are made from vitreous china, although a selection of contemporary-styled washbasins

1

made from various materials, such as stainless steel or glass **1**, is now available from certain retailers. These new materials may look very stylish and modern, but they are generally quite expensive to buy and often require more work than usual to keep clean.

Depending on the overall style of your bathroom, washbasins can be hung from a wall **2**, placed on a matching pedestal **3** or frame **4**. Washbasins can be fitted into the top of a storage unit **5**, which may be

2

freestanding or part of a run rather like in a kitchen. A freestanding corner unit is also useful if you have limited space, for instance in a cloakroom **6**. If your bathroom is large enough, consider installing a double handbasin unit **7**. This 'his-and-her' arrangement is great for busy working couples, but a large family would also make good use of it. You could have, for instance, underslung washbasins set in a granite, limestone **8** or marble top. An old pine cabinet beneath can provide attractive storage.

3

4

5

6

7

8

SHOWERS

In today's fast moving life, especially in the city, a shower is on nearly everyone's shopping list as a must-have in order to keep to our tough schedules and remain hygienic. For me a proper powerful shower is an absolute necessity. I have to agree it's not the same as a bath (apart from the obvious differences). I just wouldn't be able to manage what I do today without one, particularly as I do what is considered to be a dirty job. I always shower first thing in the morning, either after an enforced early morning workout or before an early start for work. Whichever, it is a head clearing and invigorating start to the day. A bath on the other hand is something I enjoy when I have time to lie back and relax, with a good soak to ease the ageing muscles.

However, I must mention that a poorly fitted, unsuccessfully operating shower is nothing short of totally useless. The whole principal of a shower is to wash you with pressurized water – this is what cleans and revitalises you, so in order for the shower to fulfil its purpose be very careful with your choice. There are lots of different types to

5

6

choose from and where showers are concerned, it's definitely not a case of the simplest is the best because, more often than not, creating a very successful shower can be expensive and complicated to fit. It is definitely a case of fit the best you can afford. The choice is very wide, ranging from sophisticated walk-in showers **1** to compact corner units **2** and **3**. If your shower unit can only fit into a bedroom, there are now traditional and contemporary styles available **4**. If the only space available is over the bath there is a range of glass side panels **5** to protect your bathroom There are some wonderful shower fittings which can give you that fabulous drenching shower **6** to the latest all-in-one unit with body sprays as well **7**!

7

TOILETS

The choice of toilet type for your bathroom will depend on its overall style. If the toilet is in a room of its own, to save space it makes sense to fit a toilet that fits neatly into a corner **1**.

The high-level cistern type **2** is the oldest design and suits many period styles. The cistern is mounted on the wall at high level, close to the ceiling, and connected to the pan by a long, surface-mounted flush pipe. This simple wash-down system is operated by a chain. When the cistern flushes, the force of the released water pushes the waste around the trap and down the waste pipe. Although it may be a touch noisy, I prefer this system because it is very efficient. The most common toilet in use today, however, is the standard low-level one, which relies on a siphonic flushing action. Here, the cistern is mounted on the wall directly above the pan. It is operated by either a handle fixed to the front of the cistern or a button in the top **3**. This type of toilet works efficiently and is the simplest to maintain. The third type of toilet has a concealed cistern fitted behind a false wall. The pan can either be wall-mounted **4** or a standard floor style **5** and suits many modern bathroom designs.

3

4

1

2

5

TAPS

6

7

8

9

10   **11**

You've probably heard that old saying "spoilt for choice!" – never will it be more pertinent than when choosing both taps and mixers.

You'll find there's a huge range, with hundreds of different styles and finishes. We've come a long way from the individual hot and cold taps **6** to all sorts of weird and wonderful ways of turning on and off **7**, mixing water (hot and cold) to obtain the desired temperature. As always good design must encompass both aesthetics and practicality, hence we seem to have come full circle (like most fashion!) to some of the most popular designs today which very much replicate early designs **8**. Also popular is the monobloc type; a pair of taps with integral mixer all in one unit, normally fitted through a single central tap hole in the sink, on the wall **9** and **10** or on a surface **11**.

FLOORING

What should you use for a bathroom floor? The important thing to remember about bathroom floors is that wherever water is used in abundance, there is a potential danger to any wooden floor substructure. So any floor covering that you choose needs to be waterproof in itself, but you will also have to ensure that it is well sealed all around with a suitable silicone mastic sealant.

Apart from water resistance, the flooring needs to be comfortable to walk on in bare feet, and it must look good with the rest of the decor. Personally, I think carpets never work in a bathroom that is in frequent use. Look instead for a practical, low-maintenance flooring to suit your budget: polished hardwood or Amtico **1** and **4** flooring at the top-end of the market, or a good-quality lino **2** or vinyl **3** at the less expensive end (see pages 108–109). These are available in a staggering range of colours and patterns, as tiles or in sheets for self laying or for professional installation. For the ultimate waterproof surface, you may prefer to lay ceramic tiles **5** and **6**, but make sure they are slip-resistant. As they can be chilly in winter, I would recommend you also install underfloor heating (see page 112). Normal thickness ceramic and stone floor tiles generally do not present a weight problem, but they do require a sub-floor. (see page 110).

1

2

3

4

5

6

WALL TILES

The type of tile you choose for your bathroom is very much a matter of personal taste. There's certainly no shortage of choice, so you can create virtually any look you want, especially if you mix and match sizes and colours in a well-planned design **1**.

Smaller tiles **2** take longer and are more fiddly to fit than larger ones, but mosaic tiles **3** and **4** come on large sheets that are easy to fit. Mosaics are ideal for tiling small or awkward areas, or for use as a decorative feature. Border and picture tiles (there are literally hundreds of designs) can be used sparingly to great effect **5**. Marble, granite and many other natural types of stone are also widely available in convenient tile sizes.

2

1

3

4

5

ACCESSORIES

Permanently fixed bathroom accessories like mirrors, cabinets, shelving, towel rails, soap dishes and toilet roll holders are available in a wide range of materials and colours to complement your bathroom. Whether made of plastic, glass **1**, wood or metal **2** and **3**, they are the final touch that puts the seal on its style. Metal accessories are available in a range of finishes from brass through to gold, including new and antique looks. For me though, chrome is the classic bathroom finish. As it resists tarnishing better than any other metal, it is a

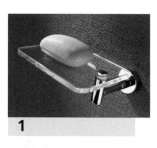

1

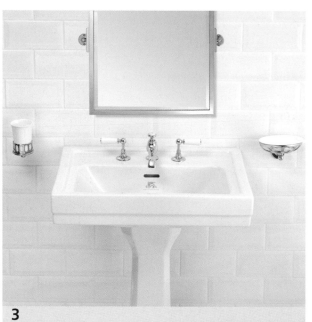

3

2

4

particularly good choice for taps, plugs, showerheads and mixers, and it's only logical to choose fixed accessories to match, even down to a funky radiator **4**. Whatever the style, a chrome finish is reliable and ageless; it stays clean and bright for years and buffs up beautifully with a soft cloth.

One of the most important bathroom accessories is the cabinet, where you can tidy away your toiletries, medicines etc. There are now some really ingenious storage ideas, from revolving cabinets that present a mirror on one side while hiding all your secrets behind slatted shutters on the other side **5**, to cabinets where the contents slide out at a touch **6**. Most cabinets now incorporate some form of lighting and some also include a shaver switch **7**. If you don't have room for a mirror and a cabinet, make sure you buy a mirror-fronted cabinet.

Storage can also be created beneath the washbasin, even a sit-on type **8**. Where space is limited, a fitted bathroom not only hides the plumbing but also provides masses of storage **9**, including around the bath **10**.

7

8

9

5

6

10

designing your bathroom

Fitting a new bathroom can become one of the most major and expensive tasks you ever undertake in your home, possibly second only to fitting a complete kitchen. It is important, therefore, to decide what you need and want from your bathroom – for both the present and in the future – and to incorporate all these requirements in your bathroom master plan.

OVERALL CONSIDERATIONS

In most house designs the bathroom is a relatively small room, which I think is totally inappropriate to the amount of use the average family bathroom gets. I think a small bathroom must be a legacy of Victorian times, when a weekly bath was considered the ultimate in personal hygiene and so only required a relatively small space.

Thankfully, we have moved on, and the modern bathroom features as one of the most important rooms in the home – and boy do I know it! I have two young teenage daughters who, although they have their own bathroom, seem to think that every other bathroom and shower in the house is off-limits to everyone else when they are bent on a session of self pampering or beautification. Like all good Dads, and husbands, I very sensibly make myself scarce at these times and retreat to the refuge of my garden shed.

I know I'm making light of it, but if your home doesn't have sufficient facilities for your family's needs, this scenario could become a serious inconvenience. So plan carefully, not just for now but for the future, when your little girls, and boys, grow up to hog the bathroom for hours on end. A second bathroom, or perhaps a new shower room, may be the answer, if you have room to fit one in your house. Although this might not guarantee the completely smooth running of the bathroom facilities in your household, it would go a long way to help.

To help you think things through, the following pages list some basic ground rules to consider when planning a new bathroom. There are also some tips on timing, budget and drawing plans.

TOMMY'S ADVICE

When planning your bathroom, there are a number of things you need to think about and plan carefully before you start work. For example, you need to consider the size of the room and whether you want to enlarge it, the kind of suite you want, and how much money you actually want to spend. Here are some points you should take into consideration:

GENERAL POINTS

♦ Is the existing bathroom big enough for future, as well as present, requirements?

♦ Ask everyone who's going to be using the bathroom on a regular basis what they would like from the new room.

♦ Could you increase the size of your bathroom? How?

♦ Could you join a separate bathroom and toilet together to create one spacious room?

♦ Could you move the door to incorporate a section of hallway, or an adjoining room to create a larger bathroom?

♦ To retain a separate toilet, could you install a new one elsewhere, for example under the stairs?

♦ What style of bathroom do you want – period, traditional or contemporary?

♦ Is there a room adjoining a bedroom that could be turned into an en-suite bathroom? Or could part of a bedroom be altered to create an en-suite bathroom or shower room?

♦ What could you fit comfortably in your bathroom? Should you sacrifice the bath for a big shower? Should you have a 'his and hers' washbasin unit instead of a bidet?

♦ What decorative finishes – tiles, glass, marble, wood and so on – do you want?

♦ What lighting and accessories do you need?

IMPORTANT CONSIDERATIONS

♦ Consider carefully the long-term effect of your choice of colour for the bathroom suite. White is easy to clean and never goes out of fashion.

♦ When choosing fittings and attachments, consider maintenance. Chrome is low maintenance and polishes up beautifully. It's also a perfect match for a white suite.

♦ Ceramic tiles completely covering all walls and floors might make the bathroom appear cold and also create condensation problems.

♦ A free-standing shower **1** is vital if you have the room. Ensure the shower is correctly fitted – damage resulting from a poorly fitted shower can be expensive and difficult to repair.

♦ Ensure the bathroom is effectively heated. A cold bathroom is unwelcoming and not conducive to a pleasant ambience. Consider installing a radiator that doubles as a towel rail **2**.

♦ Try to incorporate adequate storage in your bathroom **3**. This enables you to avoid lots of messy clutter, which would spoil the look of your new bathroom.

♦ Create attractive and interesting lighting. This will, effectively, be the icing on the cake. Done well, it will make the bathroom look and feel wonderful **4**. Seek advice on the suitability, safety wise, of any lighting designs you choose.

♦ Think through the decoration options carefully – would you like a bright and breezy bathroom or a rich, themed colour scheme. The days of slapping a coat of neutral emulsion across everything are thankfully gone, and you can now choose from a range of wonderful colours.

♦ Ensure you have adequate ventilation in order to avoid condensation problems, which could ruin the look of your new bathroom in no time at all. This may mean installing vents or an electrically assisted ventilation system.

1

2

3

4

TIMING

In order to create a successful campaign out of remodelling your bathroom, plan carefully how long it will take. You won't like not being able to bathe for very long, so if you are going to undertake the whole project yourself over a period of time, it may be appropriate to arrange for bath sharing with relatives or neighbours. I don't mean literally, of course, but I suppose that depends on who your neighbours are!

Avail yourself of all the tools and materials you'll need, and store them carefully in the spare room or somewhere convenient before you start. If you feel the whole project may be too much for you, consider breaking it down into more manageable chunks, and bring in skilled tradesmen to advise, check, or carry out any work you feel uncertain about undertaking yourself. You can save a considerable amount of money by doing a lot of the preparation work yourself even if you call in an expert; you can cap off the water supply, strip out the old bathroom suite, remove any tiles and demolish any partition walls beforehand.

BUDGET

Don't lose sight of your budget, and always allow for a 20–25 per cent overspend when working it out. When buying a new bathroom suite, beware of budget-price suites costing a few hundred pounds only. The quality is normally poor and only very basic items are included; I'm reminded of the phrase, 'You pay peanuts and you get monkeys!' Do look around – great bargains are sometimes available from quality ends of line and ex-showroom suites. Quite often, expensive design look-alikes are available from different manufacturers at a huge saving. It is also worth checking out the Internet and phoning around your local suppliers to ask if they have anything on special offer before you set out.

DRAWING UP YOUR PLAN

Once you've decided on the style of your bathroom, you can draw up your plan, taking all the points I have mentioned into consideration. Make a rough drawing of on paper **1** then measure your space accurately and mark it on the plan. Take the longest dimensions across the

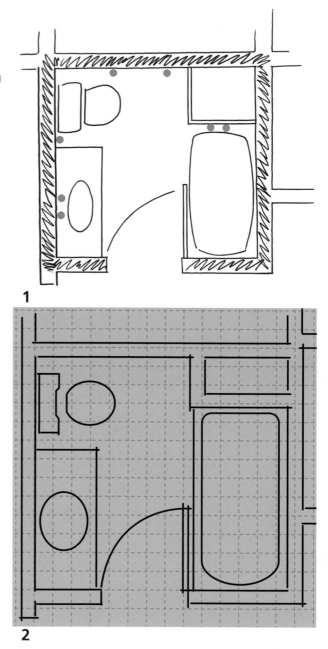

1

2

room, including every alcove and return, note the way the door opens, and mark where the window is – how high up from the floor and the distance on either side. Mark exactly where the radiator and water supply pipes are.

Now you can draw up the final plan accurately on graph paper **2**. Some bathroom suite brochures include scaled shapes that you can cut out and move around your plan. Many manufacturers and suppliers offer a planning service and will draw up detailed plans for you.

bathroom fitting

You don't have to be a bright spark to start seeing the fruits of your labour start to appear with the fitting stage. Where's that bath? I need a long soak after all this work!

fitting a bath

There are some simple and sensible rules to follow when installing a bath. If you've decided to fit a new bath in a different position from the old one, this will mean altering the pipework, which will need to be done before putting the bath in place.

INSTALLATION

When fitting a bath you have to plan the order in which you will need to work and install any necessary new pipework. Measure the bath, pipe and waste positions, and mark them clearly on the wall in pencil for easy reference. Over the years, this habit has saved me from making many a silly, time-consuming mistake.

You may need to run new 22mm (⅞in) supply pipes or add spurs to the existing runs. Do this before you actually install the bath, in readiness for connecting to the flexible pipes attached to the bath taps or mixer **1**.

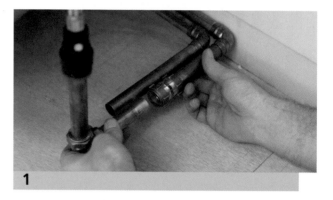

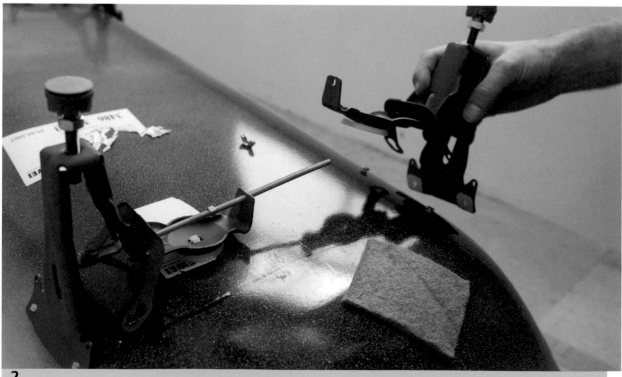

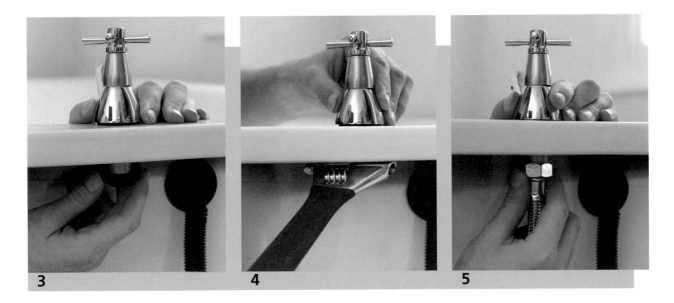

3   4   5

ATTACHING THE FEET

Before you do anything else, you need to attach the feet assembly. Turn the bath upside down; keep the bath in its packaging so it doesn't get scratched. On an old-style, cast-iron bath the feet, normally of a ball and claw design, are simply fitted onto predetermined positions using the bolts provided. A plastic bath may have a supporting frame, with legs attached, which has to be fitted before the bath is installed. On pressed-steel baths, the legs are either similarly bolted on **2**, or stuck to the base by means of an adhesive pad attached to the leg assembly. The leg positions are important, so check with the instructions that you have correctly fitted the leg assembly. You may have to adjust the leg heights to suit an uneven floor when you finally install the bath (see page 69).

FITTING THE TAPS

It is extremely difficult to fit the taps and the waste and overflow once a bath is in place, so you need to this before you finally position it. Slip a plastic or rubber sealing gasket over the tap or mixer tail, then pop this through the tap hole, so the gasket (which will ensure a waterproof seal) sits between the tap and the bath. Slip a top-hat washer over the tail **3**, then tighten the back nut onto the tail to fix the tap or mixer body to the bath **4**. Connect the flexible 22mm (⅞in) pipe tap connector **5**.

FITTING THE WASTE AND OVERFLOW

Most baths accept a combined waste-and-overflow unit. The waste is the plughole that removes the bath water, and the overflow prevents the house flooding if you leave the bath running while you have a cup of tea. There are basically two types: a compression unit, and a banjo unit. The banjo unit must have the overflow section fitted before the trap, while the compression unit fits directly to the trap itself.

To fit a banjo waste unit, first attach the overflow pipe to its inlet. Fit the washer seal over the overflow grille. Insert the threaded overflow boss from the underside of the bath through the overflow hole and screw the overflow grille onto it **6**.

6

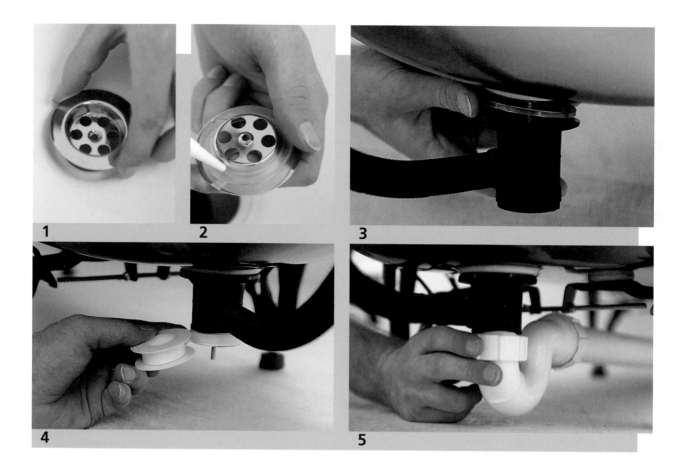

1　　　　**2**　　　　**3**

4　　　　**5**

When fitting the waste outlet, slip the rubber washer over the tail then insert it into the bath waste hole **1**.

👍 **TOP TIP Add a bead of silicone mastic sealant to the washer and waste before inserting into the waste hole 2.**

Hold the waste fitting, with its washer in place, beneath

the bath waste hole **3**, then screw the waste outlet into it. Wrap several turns of PTFE tape around the thread of the waste fitting **4**, then tighten the bath trap nut onto the threaded tail of the waste **5**.

👍 **TOP TIP To avoid damaging the chrome, wrap a cloth around the outlet grille before tightening with grips 6.**

6

ENSURING THE BATH IS LEVEL

As a guide for levelling the bath, make pencil marks along the wall with the aid of a spirit level **7**. When everything is ready, check the final position of the bath with the spirit

7

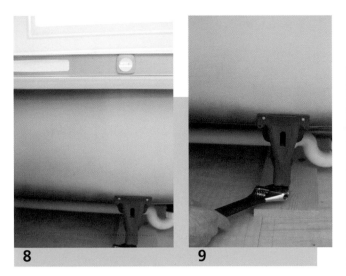

8

9

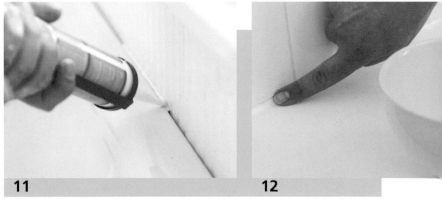

10

level along both the length and width **8**. Most height adjustments are made by turning the adjustable legs up or down **9**. Cast-iron baths don't have adjustable legs, but fine adjustment can be made using the bolts and fine washers or packers.

SUPPORTING THE BATH

Try to bear in mind the amount of weight a bath full of water would weigh, then add your own body weight. As a necessary precaution – and in order not to surprise anyone in the room below with an unannounced visit – I suggest that you strengthen the floor either by fixing 19mm (¾in) plywood under the bath, or simply fit two boards beneath the legs in order to spread the weight over a greater area of floor **10**.

The added bonus of doing this is that you reduce the movement levels of the bath between its full and empty states, which enables you to make a much more durable and effective water seal between the bath and wall.

MAKING THE SEAL

Once the movement levels between the bath and surrounding walls have been minimized by supporting the bath properly, it's time to make the waterproof seal between the bath and the adjacent walls. An effective seal is paramount to prevent damp problems occurring later on.

Ensure both surfaces are completely dust and grease free. Grease shouldn't be a problem, though, if you're fitting a new bath to new tiles. To get the right finish on the mastic, cut the nozzle to the required width of mastic. Fit the tube in the applicator and start applying from the corner, if there is one, outwards. Keeping your hand steady, move slowly but continuously in the desired direction **11**, using a clean damp cloth to wipe the sealant. While the mastic is fresh, dip a finger in some soapy water and run it slowly over the mastic to effect a smooth shape and ensure contact with both surfaces **12**. A small bowl of equal quantities of washing-up liquid and water is all you need. Alternatively, try using the handle of a fork or teaspoon to shape the mastic. Allow the mastic to dry for at least 24 hours before using the bath.

It's important to use good-quality silicone mastic sealant, as it will incorporate essential ingredients, such as elasticity, colour retention and anti-fungal inhibitors.

11

12

fitting washbasins

Basins can be hung on a wall or placed on a pedestal or, as is becoming increasingly common, fitted into a unit in your bathroom, in a run like a kitchen. Some basins are simply placed on top of a unit, to look unfitted.

INSTALLING A PEDESTAL WASHBASIN

The most commonly used washbasin is the pedestal type. This requires the basin to be secured to the wall while resting on a matching china pedestal, which provides balanced central support. The pedestal incorporates a void for hiding the supply pipes and waste.

Pedestal washbasins are available in lots of wonderful designs and different colours. There is also a wide choice of taps and mixers with plugs or pop-up wastes, so remember to purchase the taps and waste at the same time as the basin, to ensure you get exactly what you want. It makes sense to fit the taps before you install the basin, after which access will be awkward to say the least.

If you are simply replacing an existing pedestal basin, the process is relatively simple. Mark the position of the basin on the wall, using the pedestal to position it **1**. Screw the basin to the wall **2**, or hang it on brackets depending on the type. Slide the pedestal underneath **3** and connect the supply pipes to the taps **4**. Fit the trap to the basin waste

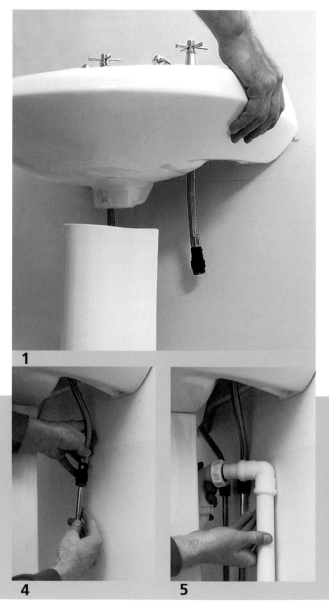

5. Screw in the plughole **6** and attach the plug. Make a mastic seal around the pedestal top and the basin base **7**. Once you have done all your tiling and flooring, you can seal where the basin is in contact with the wall and around the base of the pedestal (see page 69).

Sometimes, things are a little less straightforward. Some new mixer taps won't attach to old tap connectors, and require simple pipework alterations (see page 18). If you are re-siting a pedestal basin, you may have to alter and conceal the supply pipework and waste pipe. These could be hidden behind the skirting if the wall is studwork, or chased in if the wall is solid. Ensure that rodding eyes for servicing are fitted to both ends of the waste pipe run. Remember, too, that the waste pipe requires a slight fall – a minimum of 6mm (¼in) to every 300mm (12in) of pipe run. The total pipe run length should not exceed 3m (approximately 10ft).

New pipe runs are best fitted to the wall surface, then boxed in (see page 101). Take care not to damage the pipes when fitting, and ensure it is detachable for access.

FIXING TO THE WALL

If you are attaching a new washbasin to a solid wall, make sure the screw fixings or mounting brackets are properly anchored with plugs. If attaching to a hollow stud wall, the screw positions may not align with the studs. In this case cut out the plasterboard **8** and screw a softwood batten or piece of plywood to the studs **9** to hold the basin.

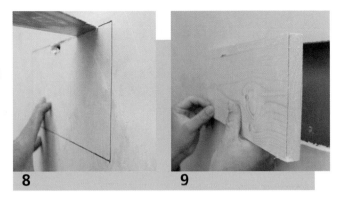

8   **9**

👍 **TOP TIP To make sure everything will fit perfectly, it pays to temporarily set up your basin and run all your pipework. Then you can disconnect the basin for any wall repairs and floor covering to be completed, before making the final connections and sealing around the pedestal base.**

6   **7**

INSTALLING A WALL-HUNG BASIN

As they are usually quite small, wall-hung basins are particularly useful for a small space, such as a separate toilet. Some wall-hung basins are designed to fit into the wall itself, but the plumbing involved with these can be a bit complex. Most wall-hung basins are fitted onto a concealed wall-mounting bracket, which must be fixed to the timber wall studs or a softwood batten attached to the studwork, or to brick- or blockwork. Concealing the pipework may involve running the pipes in the studwork or chasing them into the brickwork.

Alternatively, the basin could be set against a false wall, in which the plumbing is hidden. A chrome waste trap would look better than a plastic one. Or you could conceal the plumbing in a purpose-made cabinet (see page 80).

The waste pipe needs to be less than 2m (6ft 6in) long, otherwise the water in the trap could be siphoned out, allowing unpleasant smells from the drain to enter your bathroom. (In the trade, this is known as 'pulling the trap'.) If the run has to be longer than this, you'll need to fit an anti-siphon trap or use larger waste pipe.

For most people, a wall-hung basin needs to be fixed 800mm (2ft 8in) above the floor, with a minimum 1.1m (3ft 8in) clear width to allow for elbow room when washing. My wife hasn't noticed that our basin is 100mm (4in) higher than everyone else's – I've got to think of my poor old back!

Mark the position for the basin on the wall **1**. If necessary, fix an extra noggin at the back of the wall to support the brackets. Position the pipes at the back and through the plasterboard. On the front of the wall draw a level pencil line where the brackets are to be fixed corresponding with the noggin at the back. Screw the brackets in place **2**. Attach the taps and waste pipe to the basin **3** (see also page 70). Hook the basin onto the brackets **4**, then screw the basin to

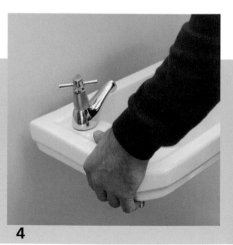

2

1

3

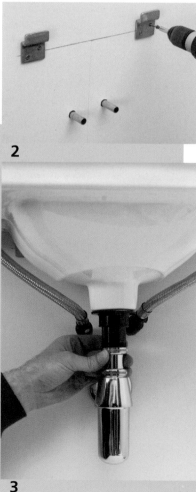

4

6

7

the wall through the holes on the underside. Hold the waste assembly in place and fit (see page 67). Now enjoy **5**.

Where space is especially tight, such as in a cloakroom that makes the most of limited floor space, a well-designed wall-hung corner basin might be the answer. Some designs incorporate a small shelf and even a hanging rail **6**. Corner basins are fixed in the same way as ordinary wall-hung basins. Another possibility is a wall-hung basin with an integral unit **7**, which allows for the waste pipework to be routed through the wall, rather than having to cut holes in the unit. Also, instead of needing to buy an expensive chrome waste fitting, you can simply slide a U-shaped metal sleeve over the plastic pipework, which is also an added support for the basin.

5

fitting toilets & bidets

Excuse the pun, but before you become too flushed about a successful purchase, remember you still have to fit it! Try to replace the pan in the original position. Keeping it against an outside wall near the soil stack will save you having to reroute the supply pipework and waste run.

PUTTING IT ALL TOGETHER

Nearly all modern toilets are close-coupled, that is with a low-level cistern sitting on the pan. They are operated by a siphonic action, which is created by a built-in single or double trap. This makes the toilet flush a lot more quietly than the wash-down type with a separate cistern.

FIRST THE CISTERN

All new toilets are supplied with the internal flushing mechanism already fitted together as one piece, so all you have to do is pop it in the cistern and attach the flush

THEN THE PAN

Fit the plastic push-fit connector onto the pan outlet **3**. Attach the cistern temporarily to the pan **4** and position the pan with the cistern against the wall **5** ready to mark the fixing positions on the floor. If the floor is a concrete screed or tiled, you will have to remove the pan and drill and plug the holes for the screws. Use non-corrosive screws, such as brass ones.

Remove the cistern and check with a spirit level that the pan is level; use strips of wood or plastic as packers if

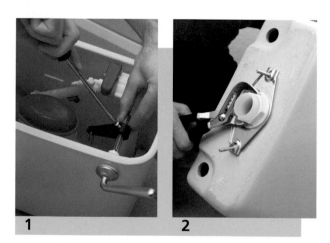

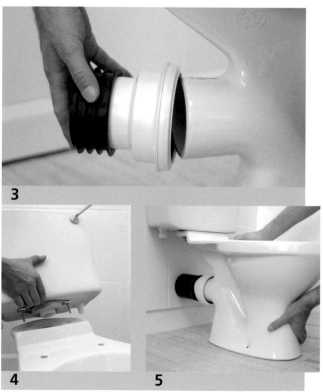

handle. Push the threaded end of the water supply pipe through the hole in the base or at the top of the cistern and secure it on the outside with a washer and collar. Tighten by hand, taking care not to cross-thread the collar. Fit the flush handle with a threaded collar and link it to the flushing mechanism **1**. Secure the internal flushing mechanism to the base of the cistern with a connecting plate and tighten **2**. Insert bolts either side of the plate.

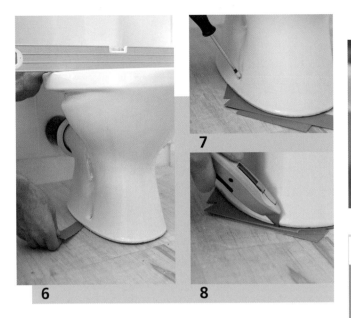

6

7

8

👍 **TOP TIP** Wrap lots of PTFE tape around fixing screws to act as a packer. I use long brass mirror screws, which can be finished off with chrome dome caps for effect.

it isn't **6**. Bed the pan on silicone mastic and screw the fixing screws home **7**. Trim any packers flush using a craft knife **8** and clean off excess mastic with a damp cloth.

If the toilet has a flush pipe, connect it to the pan. Hold the cistern against the wall to mark the fixing holes. Fix the cistern to the wall with non-corrosive screws and washers **9**, drilling and plugging the holes if the wall is solid. Tighten the flush pipe connecting nut to the cistern and connect the 15mm (⅝in) water supply pipe to the cistern float valve with a tap connector **10**.

The last thing to do is to fix the toilet seat, via the pre-machined holes in the pan and the seat assembly kit **11**. I always smear a bit of Vaseline or silicone grease on the fixing kit components before completing this task.

TOMMY'S ADVICE

It's very important that you allow for connecting an overflow pipe in case of valve failure. This is done by connecting a 22mm (⅞in) overflow pipe to the cistern connector provided, through an external wall so that it can discharge outside in the event of a problem. Ensure a slight downwards fall to the overflow pipe. If you're unable to run the overflow pipe through an outside wall, connect it to a bathroom waste pipe. A special purpose-made fitting called a Tundish (above) can be fitted to detect when water overflows from the cistern.

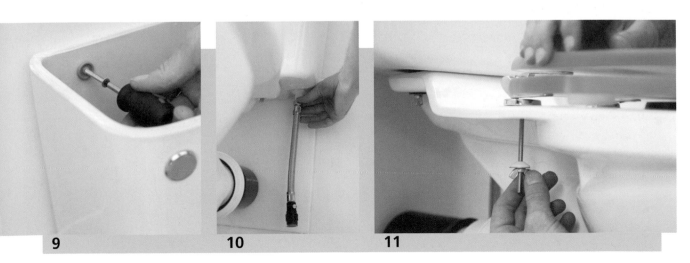

9

10

11

INSTALLING A PUMP SHREDDER UNIT/MACERATOR

If you wish to install an additional toilet in your home away from the house soil stack, a small-bore waste system may be the answer. Traditionally known in the trade as a 'chewer and spitter', this is a pump-and-shredder unit that macerates toilet waste and pushes it through a small diameter pipe, over a considerable distance if necessary.

This system allows you to install a toilet up to 100m (350ft) away from the soil stack, even pumping waste vertically to a maximum of 6m (nearly 20ft). Its flexibility means you can site a secondary toilet almost anywhere in the house – in a loft or basement conversion for instance or even under the stairs.

The macerator is supplied as a free-standing unit that fits directly behind the pan and below the cistern. It must be wired into an unswitched fused spur outlet. The site needs adequate ventilation, but this can be provided for mechanically if necessary.

The unit's waste pipe is connected to the 100mm (4in) soil stack via a standard 32mm (1¼in) strap-on pipe boss, but the manufacturer normally supplies a 22–32mm (⅞–1¼in) adaptor in case smaller gauge pipework is used. The connection to the soil stack must be made at least 200mm (8in) above or below any other connections.

If you live in a block of flats you may need approval from the management company to install a macerator,

1

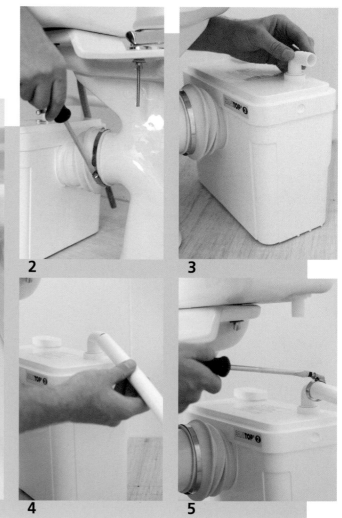

2

3

4

5

and even if you live in a house you should check with your local water authority that there are no restrictions.

The unit can be a little noisy, especially on bare floorboards. It switches on automatically when the toilet is flushed and runs for about 10–20 seconds, so some carpeting or matting underneath it might be necessary.

CONNECTING A MACERATOR

Slip a large Jubilee clip over the pan outlet then fit the flexible socket of the unit over the pan outlet **1**. Position the Jubilee clip over the socket and tighten it using a screwdriver to make a watertight seal **2**.

Next, connect the discharge pipework. Insert the discharge elbow in the lid of the macerator until the lugs engage then turn it clockwise **3**. Fit the flexible hose supplied with the unit over the elbow **4** and secure it with a Jubilee clip **5**. Use another Jubilee clip to connect the other end of the hose to a copper or UPVC waste pipe run to the soil stack **6**. You can use 22mm (⅞) diameter pipe, but any discharge pipework longer than 12 metres needs to be increased to 32mm (1¼in).

Horizontal waste pipes must have a minimum fall of 1:200 (5mm per metre/¼in per yard) **7**. If the discharge pipework runs to a level considerably lower than the macerator unit, the resulting siphoning effect can suck out the water seal in the unit. To prevent this, fit an air-admittance valve at the high point of the pipe run.

Finally, connect the discharge pipework to the existing soil stack using a strap-on boss **8** and **9**. Some are self-cutting, but you may need to cut a hole yourself with a special drill attachment. Lag any external pipework.

ELECTRICAL INSTALLATION

Ask a qualified electrician to connect the unit to an unswitched fused wiring connector protected with a 5 amp fuse, or a circuit breaker set to 30mA.

SWITCHING IT ON

When all your plumbing and electrical connections have been made, flush once. The motor should run for about 10–20 seconds before switching off. Now you can check that all the seals and connectors are watertight.

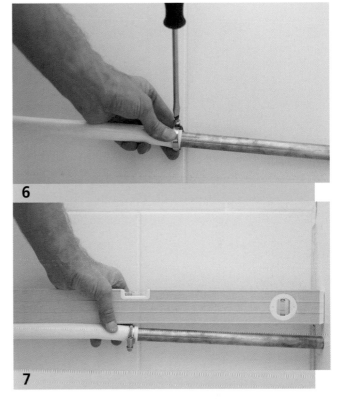

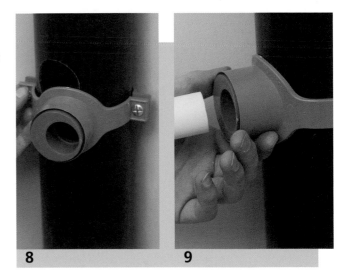

👍 **TOP TIP Only human waste and toilet paper should be placed in a macerator unit. Bulky, hard or fabric objects will damage the motor and cause blockages, which will then need to be cleared manually. Don't use any chemicals other than standard cleaners.**

CHOOSING AND INSTALLING A WALL-HUNG TOILET

In a fitted bathroom, the toilet cistern and plumbing are concealed behind a false wall, while the pan is positioned against it. As these cisterns do not need to look attractive, they are relatively cheap to buy. The plumbing is basically the same as for any other system (see page 75), except that the flushing handle or button is mounted on the wall.

👍 **TOP TIP Ensure that any concealment wall or panelling is fitted with full access for servicing and maintenance.**

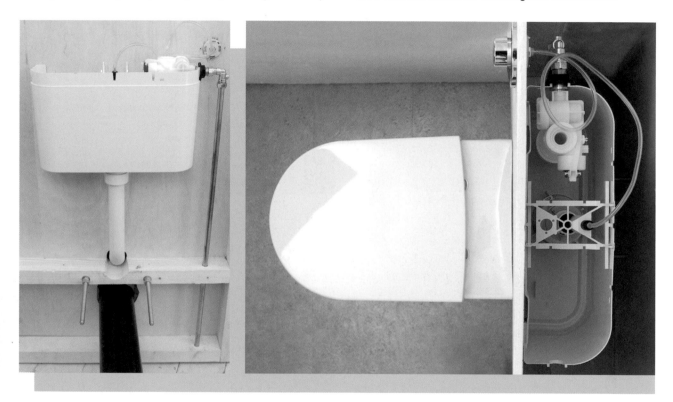

CHOOSING AND INSTALLING A BIDET

Whenever I see a bidet, I'm reminded of a well-known movie, Crocodile Dundee, where Paul Hogan, as Croc Dundee, encounters a bidet for the first time in a very swish New York hotel. So what does he do when he sees it? Like any sensible bloke, he washes his socks in it!

There are two types of bidet: one is known as the over-rim supply bidet, the other is called the through-rim supply bidet. The latter is quite a difficult piece of equipment to install, and local water authorities have strict regulations regarding the installation of this type of bidet.

So unless you're extremely proficient at DIY, this particular job is probably best left to a professional plumber. When siting a bidet, remember to allow for enough leg room – a minimum of 760mm (2ft 6in) overall.

FITTING AN OVER-RIM SUPPLY BIDET

Unlike the rim supply bidet, this over-rim type is simplicity itself to install. If you are using separate hot and cold taps, follow the steps for installing a bath (see page 66). The majority of modern bidets are supplied with a single hole for a mixer tap with a pop-up waste attached to the tap. The hot and cold supply pipes are simply branched off the existing bathroom plumbing, while the waste pipe discharges like other bathroom fittings into the 110mm (4¼in) soil pipe or external hopper.

Push the top of the pop-up waste down through the waste hole in the bidet, seating it on a layer of sealant. From underneath, screw the bottom of the waste onto the top **1**. Position the washer at the base of the tap **2**. Fit shut-off valves to the hot and cold supply to the tap. Secure the tap to the bidet **3**. Screw the pop-up waste lever by hand into the waste outlet **4**. To insert the pop-up rod, join it with the supplied clamp and screw it tight **5**. Adjust the waste bung if necessary **6**. Move the bidet into position and connect it to the supply pipes and waste pipe **7**. Secure the bidet to the floor and wall in the same way as for a toilet (see page 74).

1

2

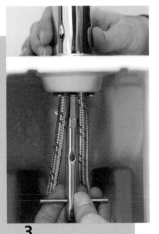

3

4

5

6

7

flat-pack bathrooms

Like fitted kitchens, fitted bathrooms look good and make use of otherwise wasted space for all kinds of handy storage. There is an enormous range available, with widely varying prices to match. The cheapest option is to put one together yourself using flat-packs.

INSTALLING A FLAT-PACK BASE CABINET

Flat-pack systems are no longer just for kitchens, and it is now possible to create a modern, streamlined look in your bathroom. The cam-and-stud fixing mechanism has revolutionized flat-packs. You simply lock the stud solidly in place by turning the cam with a screwdriver.

👍 TOP TIP **Open flat-pack packaging very carefully and store all the components together until you are ready to install. It is not at all unknown for the wrong unit to be sent out, and it will be much easier to change if everything is intact and in the original packaging.**

For a workbench, you can set a sheet of 19mm (¾in) plywood or MDF on a pair of saw horses or stools. Unpack and identify all the parts before you begin work. I always position the parts on the bench like an exploded drawing, and then check which fixings go where. Also, read the instructions a couple of times to absorb all the information.

Familiarize yourself with all the components **1**. To help you understand the construction properly and avoid making errors, carry out a dry run by putting the unit together without any fixings.

1

2

3

4

PUTTING THE BASE TOGETHER

Take the two side panels and identify the pre-drilled cam stud holes. Insert the cam studs in the pre-drilled holes and tighten these using a screwdriver; double check they are in the correct holes **2**. Take the base panel and horizontal struts and identify the pre-drilled dowel holes. The dowels should fit snugly, and act as locating pins for the connecting panels. With the unit on its side, attach the side panels to the pre-dowelled base panel and horizontal struts. Pop the cams into the holes, ensuring that the embossed arrow on the cams points towards the end of the panel. To secure the side panels to the base, tighten the locking cams; turn them clockwise using a screwdriver.

Attach the legs to the base **3**, following the manufacturer's instructions. Legs can be adjusted individually to accommodate any unevenness in the floor.

The cupboard shelves and doors are fitted once the whole bathroom is assembled and in position. Two-door units are assembled in the same way as single-door units, but sometimes with the addition of a vertical section, which is screwed to the centre of the unit front edge. This piece is added to strengthen the unit and conceal any gap between the doors.

HINGES

Open the packaging for the door hinges carefully so as not to lose any parts, and leave them in position on the workbench while you work on the door. This will reduce the risk of damage as you work. First, attach the hinges to the pre-machined positions on the door. There should be

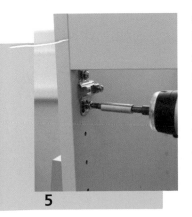

two tiny indentations, one either side of the cut-out hole, indicating where the screws should go. Carefully drill pilot holes, using a very fine bit; make sure you don't go right through to the face and ruin the door. Alternatively, you could use a bradawl to create shallow guide holes for the screws. Push the circular part of the

5

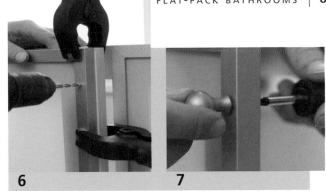

6   **7**

hinge into the cut-out hole **4**, and drive home the screws. Fix the hinge-mounting plates to the pre-drilled positions on the side panel of the unit **5**, but don't overtighten.

HANDLES

The door handle positions should be visible on the opposite side to the hinge; look for one or two tiny indentations. Hold a scrap block of wood against the face of the door **6**, to make sure breakout doesn't occur, then drill pilot holes. Align the door handles and screw the bolts in from the back of the door **7**. Squeeze an extra quarter-turn on the screwdriver, to create a tight fit and prevent the handles falling off after the first ten minutes!

installing showers

In today's fast moving world, especially in the city, a shower is on nearly everyone's shopping list as a must-have in order to remain hygienic while keeping to our tough schedules. For me a proper, powerful shower is an absolute necessity, although I have to admit that it's not the same as a nice, long, hot bath.

SHOWER PERFORMANCE

For a conventional shower fed by an indirect water system (see page 14) to work satisfactorily, the cold water storage tank should be at least 900mm (3ft) above the shower head. If your shower performance is poor, you can increase the water pressure by either raising the storage tank or installing a pump.

water pressure and flow are the key to a good shower

MANUAL SHOWER MIXERS

Manual shower mixers are simple to work and relatively easy to install, either over an existing bath (see page 87), or in a separate shower enclosure. They must be independently supplied with hot and cold water. The single lever controls both the temperature and flow of the water. Ceramic disc mixers operate more smoothly and are less prone to hard-water scaling. Like most shower mixers, manual mixers can be surface mounted (onto the wall face) with chromed pipework showing, or flush-fitted with the pipework and working parts of the mixer concealed within the wall, if it is a stud partition.

THERMOSTATIC MIXERS

Thermostatic mixers **1** are basically the same as manual mixers except that they have an in-built safety device to prevent scalding or freezing, if there is a change in the flow rate. Whether it's the cold or the hot water flow rate that drops, the mixer will automatically compensate, by reducing the flow rate on the opposite side. This effectively prevents an accident if somebody turns on a tap elsewhere

in the house or flushes the loo, or if a programmed washing machine, or dishwasher comes on.

👍 **TOP TIP Because of this built-in safety aspect, when fitting a thermostatic valve you need only branch off the existing bathroom supply pipes, but it's more effective to connect the pipes as near to the water tank and hot-water cylinder as possible. Bear in mind that the mixer is unable to boost the pressure or the supply, so if your pressure is low, a booster pump may be required.**

1

The majority of thermostatic mixers can be used with existing gravity-fed hot and cold water supplies, but again you may need to fit a booster pump. Check with your supplier as some showers work well at low pressure while others certainly don't.

PUMP-ASSISTED SHOWERS

A pump-assisted shower, or power shower as people commonly like to call it, appears on many a wish list as the ideal shower system. 'Power shower' does have a nice ring to it, doesn't it?

The pump delivers the water at a consistent flow rate and pressure, eliminating the minimum 900mm (3ft) water storage requirement (that is above the shower head) for a gravity-fed shower. A head, or height, of only 75–225mm (3–9in) is required to start the pump when it is switched on. A pump can be used to increase the pressure of stored hot and cold water, not mains-fed water.

The best way is to connect the cold supply directly to the cold water storage tank, not a branch from the bathroom supplies, and the hot supply should be connected directly to the cylinder via a Surrey or Essex flange, which prevents the pump from sucking air in from the vent pipe. If the hot water is heated by an immersion heater (see diagram), ensure the cylinder is supplied by a direct cold feed, and the gate valve is fully open. This prevents the cylinder from running dry, and burning out your heating element. Alternatively, if the cylinder is heated via the boiler, ensure a thermostat is fitted to prevent the water from getting too hot, making the shower splutter.

Power showers often have an integral electric pump in the mixer cabinet, which is fitted within the cubicle. Other pumps are made for isolated installation **2**, with hot and cold pipes going first to the pump, then onto the shower mixer. These pumps can often be added to an existing installation to improve a poorly performing shower. The best position for these pumps, if possible, is next to the hot-water cylinder, normally in the airing cupboard. As these pumps are not silent, you may want to insulate the cupboard for noise. When installing the pump, remember to install it at the bottom or below the cylinder to ensure it

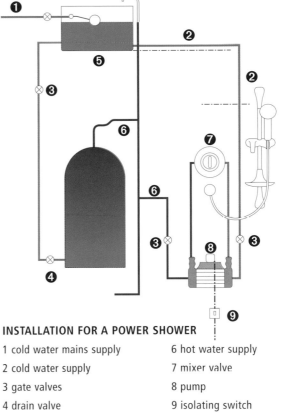

INSTALLATION FOR A POWER SHOWER

1 cold water mains supply	6 hot water supply
2 cold water supply	7 mixer valve
3 gate valves	8 pump
4 drain valve	9 isolating switch
5 cold-water storage tank	

always remains full of water. If this isn't practical, there are pumps specifically designed to operate above the cylinder at high level, even in the loft. Check with your supplier to confirm what you require.

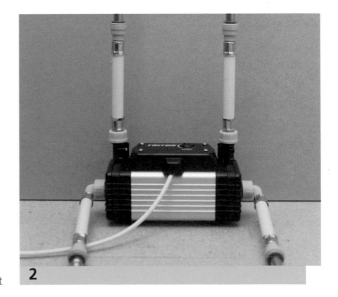

2

INSTALLING A PURPOSE-BUILT SHOWER

1

2

Probably the most efficient and attractive shower for the average home is a purpose-built shower enclosure **1**. The ideal location is a corner of the bathroom, where two sides are already formed so you only need to install a corner unit, or build a third wall and fit a glass door across the opening. From a practical point of view, I prefer to build a third block or stud wall and install a vitreous china shower tray. I can then cover all three walls with light coloured tiles that reflect as much light as possible.

There is no easy answer to the cleaning question. To keep everything spotless and sparkling, the tiles, grouting, and mastic all require regular maintenance.

👍 **TOP TIP There are lots of shower door designs, but it makes sense to select a pivot door of some kind, which, when opened, will allow the water to run down inside the enclosure 2, rather than onto the bathroom floor.**

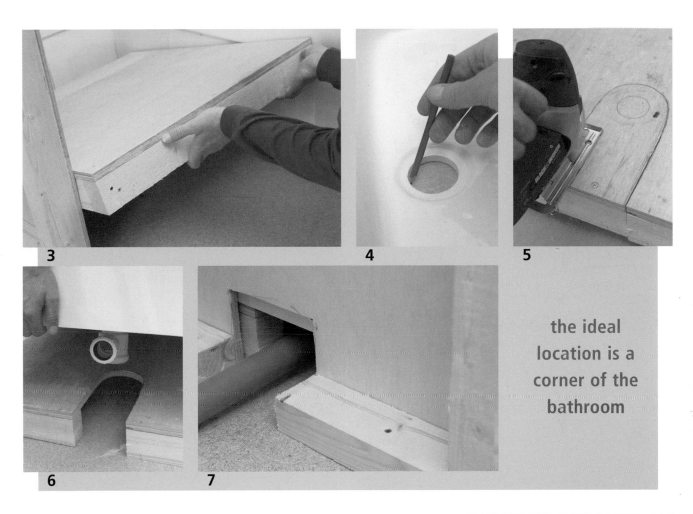

3

4

5

6

7

the ideal
location is a
corner of the
bathroom

Once the wall is built, make a solid, perfectly level base for the shower tray to sit on **3**. Use 75 x 50mm (3 x 2in) timber and 19mm (¾in) exterior-grade plywood. Place the shower tray on the base and mark out the plughole **4**. Remove the tray. Mark out a section on the plywood to take the trap and waste pipe, then, using a jigsaw, cut this out **5**. You can now fit the trap and waste pipe to the shower tray so that it will sit in the void area below it **6**. The shower waste can exit through the new wall **7**, and be concealed behind a false skirting, which will allow for access if necessary. Fit rodding eyes to the waste pipe so that any blockages can be cleared easily.

I always cut a second piece of plywood to protect the fitted shower tray from possible damage as I tile. To allow space for tiling right down to the shower tray, make the protective cover a loose fit **8**.

8

1

2

4

3

6

5

👍 **TOP TIP** Tape over the open tails to avoid any debris blocking the pipes.

You can now tile the walls (see page 102). Once the grouting and sealant have been applied (see pages 39, 102 and 103), assemble the shower mixer **5** and head. Large shower heads, the ones that give you a real drenching, are secured through the wall and attached to the water pipes by means of a tap connector **6**. Finally, fix and hang your choice of door (below), or a curtain.

A stud wall allows for the pipework to be concealed easily within it. The trick is not to clad the rear face of the wall until all the pipework is complete. Use the template supplied with the shower mixer or draw round the faceplate **1**, then cut a hole 12mm (½in) within the outline. Insert the mixer bracket through the hole and fix it in place with the screws supplied **2**. For a solid brick- or blockwork wall, you will need to chase out channels to conceal the pipework. Insert the mixer unit and screw it to the bracket **3**. Connect the hot and cold water pipes to the mixer unit **4**. Fix a copper pipe to the top of the mixer unit to take the water to the shower head. For most shower heads, it will need to exit through the wall via an elbow fitting, with the tail protruding into the shower enclosure approximately 100mm (4in) from the wall face.

INSTALLING A SCREENED AREA FOR A BATH

If you don't have room for a shower unit, you can install a shower over your bath or fit a mixer tap with a shower attachment, but you will need to fit a waterproof barrier. A purpose-made bath screen or a shower curtain will protect the rest of the bathroom from the spray.

FITTING A SHOWER CURTAIN

If you are using a bath mixer to shower with, a low-cost solution is to fit a shower rail and curtain. Fix the shower head holder at the right height, then screw the shower curtain rail brackets to the wall higher than this by drilling into the tiled area, taking care not damage the tiles (see page 121). Secure the brackets to each wall with the screws supplied (you'll need to plug a solid wall).**1**. Attach the curtain to the rings – voilà an instant shower screen **2**.

FITTING A FOLDING BATH SCREEN

A bath screen that folds is useful as you can push it out of the way when you are not showering. The hinged screen comes ready to fix to the wall, with rubber sealing strips at the bottom already fitted. Follow the instructions for fitting to the wall above. Slot in the bath screen then unfold it. As you unfold, the rubber seal should fit snugly against the bath edge. You can start leak-free showering straightaway **3**.

1   2   3

FITTING AN ELECTRIC SHOWER IN A CORNER UNIT

1

Combining a corner shower surround with an electric unit makes for an instant shower in more than one sense, as you don't need to build an extra wall to make the enclosure. The main advantage of a clear-glass surround is that it looks wonderful while letting in lots of light **1**. The big disadvantage, though, is that it may take a bit of effort to keep it sparkling, especially if you live in a hard water area and you don't have a water softener. If you squeegee the glass off after each shower, however, and train the rest of the family to do the same, you can reduce the major cleaning work considerably.

FITTING THE ELECTRIC UNIT

Electric showers have definite advantages and disadvantages. They are relatively cheap to buy, and fairly simple to install. Requiring only a cold supply direct from the rising main, there is no need for cold water storage. The water is heated instantly, as it is forced through the shower unit's heating element under mains water pressure. A built-in, thermostatically controlled safety valve controls the temperature, and will switch itself off automatically if the water pressure is too low. Ensure that a non-return valve and an isolating valve are fitted to the supply pipe to allow for easy servicing (see diagram below) or unit replacement, should the need arise.

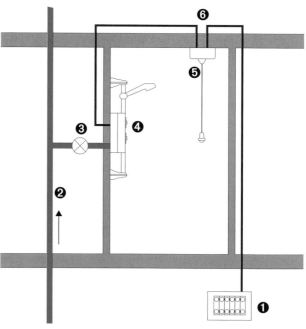

WATER AND ELECTRICITY SUPPLY

1 separate circuit from consumer unit
2 mains water supply
3 non-return valve
4 shower unit

5 ceiling-mounted double-pole isolating switch
6 mains electric supply (via double-pole switch)

3   4   5   6   7

The flow of water through the shower unit triggers a switch to turn on the element that heats the water as it passes through. Because there's so little time to heat the flowing water, a powerful electrical load is required, anything from 6–10.8 KW. This must be supplied by a separate radial circuit, protected by a 30 milliamp RCD **2**.

The electric supply cable must be a minimum 10mm^2 twin (two-core) and earth cable. The circuit needs a 45 amp double-pole isolating switch, which is usually a ceiling-mounted type as shown in the diagram opposite. The switch must have a mechanical on-off indicator as well as a neon one. It is also imperative that the shower unit and all metal pipes are bonded to earth (see page 24). If you are in any doubt about the wiring requirements ask a qualified electrician to advise you. Indeed, you will almost certainly need to get an electrician to bring the new circuit for the shower from the consumer unit.

Once the electricity supply has been installed, fit a single 15mm (⅝in) pipe to bring water direct from the cold supply to the wall. Fix the shower unit to the tiled wall (see page 121 for fixing to tiles). With the power still switched off, feed the electric cable through the backplate of the unit and wire it up **3**, following the manufacturer's instructions. Join the water inlet pipe to the unit and tighten with a compression fitting **4**. Fit the cover and attach the shower hose to the outlet at the base **5**. Attach the shower head to the wall, high enough to avoid any back-siphonage **6**.
👍 **TOP TIP Buy a shower waste with a pull-out section so you can unclog the waste overflow regularly before a build-up of hair can cause blockage problems 7.**

SHOWERING ALTERNATIVES

Ideally, my next bathroom will be a wet room without a door or shower tray, where the water will run off a water-proofed stone floor into an outlet in the floor, which, when it needed cleaning, could be done with a hose and a long-handled squeegee in about five minutes flat.

There is a downside to this dream, though – it will mean 'tanking' the room in order to make it absolutely watertight, otherwise I might be showering in my living room! Tanking is a job for a professional, however. So if you share my dream, but are worried about water going everywhere, it might be a better idea to instead fit a glass panel to enclose your showering area **1**.

Some of the latest showers have taken the shower cubicle to new heights. Some even have the shower fittings attached in one piece to one of the columns **2**, so fitting the shower becomes a real doddle **3**. All you have to do is attach the hot and cold supply to the appropriate pipes (that is after you have done all the hard graft of ensuring the pipes and drainage are in the right position – see pages 18–21). Other designs have evolved into a complete walk-in enclosure **4**, which creates the illusion of a wet room, without any doors to close behind you.

1

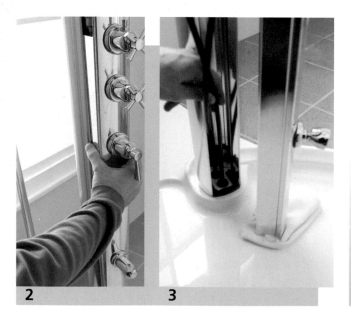

2   3

4

bathroom surfaces

Think about bathroom surfaces: easy to clean, non-slip, a thing of beauty! No, not me, although I have been called slippery in the past! Choose those surfaces carefully because they must complement your fittings.

panelling the bath

Unless you have the rolltop cast-iron variety, your bath will need to be fitted with a panel to hide the ugly void beneath it. You can either buy a panel off the shelf or make one yourself to complement the decor of your bathroom. An interesting panel can make all the difference.

MAKING AND FITTING A BATH PANEL

The days of a painted hardboard panel fitted with a chrome-coated corner piece to cover the joint are long gone – thank goodness! Today, the list of materials and styles you can choose from is as long as your arm, so you can select one that suits both your taste and pocket.

Bath panels are available as standard stock items in plastic **1**, wood in various finishes such as mahogany, new or old pine, limed oak etc, and even granite. A new idea is to incorporate cupboards and shelving in what would otherwise be wasted space **2**.

Alternatively, you can make your own panel. Making one from a substrate like plywood, blockwood or MDF is a straightforward job. For instance, you can make a really good-looking panel from a sheet of waterproof or water-resistant MDF, plus some skirting to match the rest of your bathroom. Adding moulding (see page 94) will make it more decorative. For a more creative approach, plenty of exciting materials can be used imaginatively. These include copper, lead, stainless steel, carpet and other textured materials, marble, tiles and mirror.

If you do make your own bath panel, you will first have to build and fit a proper frame to attach it to. Make sure that, once fitted, the panels are easily removable in case you need to access the plumbing.

1

2

3

4

5

6

MAKING A BATH PANEL FRAME

Normally made of 50 x 50mm (2 x 2in) sawn or planed softwood, a bath panel frame fits tightly between the lip of the bath and the floor. To calculate the frame size, measure the height and width of the area below the bath. Bear in mind that the frame's position must allow for the thickness of the panel material, which generally shouldn't protrude beyond the vertical lip of the bath. This allows for the panel top to be sealed with mastic to prevent the ingress of water. Depending on the material, the panel can be fixed to the frame using face fixings or secret (concealed) fixings, but remember to ensure access to the plumbing. Construct the frame as a ladder frame. For the long side, screw the four outside pieces together, then add uprights for strength and rigidity **3**. Screw a three-sided end frame to the wall and offer the side frame up to it **4**. Screw the side frame to the end frame **5** and to the wall at the other end. Screw the frame to the floor **6**.

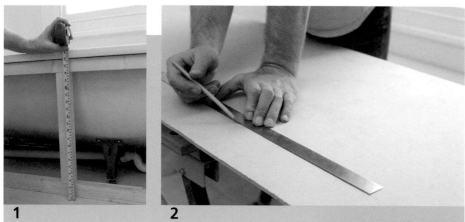

1

2

double check
that all
measurements
for the panel
and skirting are
correct before
you cut

MAKING AND FITTING A PANEL

Measure the frame area **1** and transfer your measurements to the panel material **2**, here it is a sheet of MDF. You can use a jigsaw to cut out the panel – if you do, cut a fraction outside the line, then smooth off the excess using a hand plane **3** for a neat finish.

Next, fix the skirting to the panel pieces, having first mitred the corner ends. Use PVA glue and pins if attaching

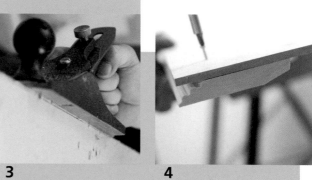

3

4

5

from the outside, or screw from the inside **4**. To add mouldings, draw a line of equal margin all around the panel. Cut and mitre the mouldings then glue and pin it along the lines. Finally, paint the panels (including the edges) with a coat of primer, two undercoats and a topcoat **5**.

FITTING A READY-MADE BATH PANEL

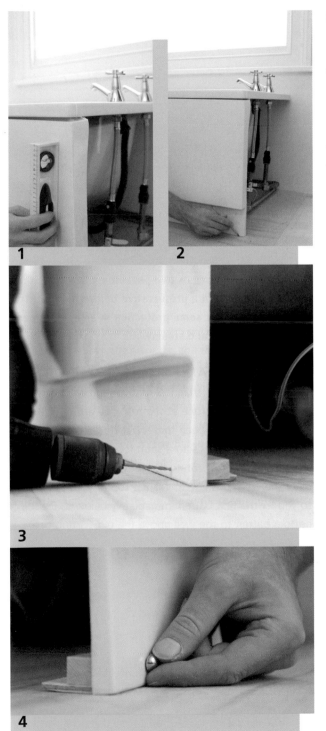

Position the panels vertically, using a spirit level or plumbline **1**, then mark a line on the floor along the length of both the panels **2**. The end panel fits behind the front panel at the corner joint. Remove the panels, and mark a second line 3mm (⅛in) inside the first. Screw battens to the floor along and inside this second line – a minimum of three battens for the front panel and two for the end one. For the batten to take the bottom flange, or turned-in edge, of the panels, you need to make a rebate. You can form this by placing a strip of packing the same depth as the flange beneath each batten. The packing will butt against the flange, while the batten above it will butt up to the inside of the panel itself. Reposition the panels, then drill 4mm (³⁄₁₆in) pilot holes through each one into the battens **3**. Secure the panels to the battens using decorative mirror screws **4**.

👍 **TOP TIP If you have a router, cut a rebate in each batten to take the panel flange instead of using packing.**

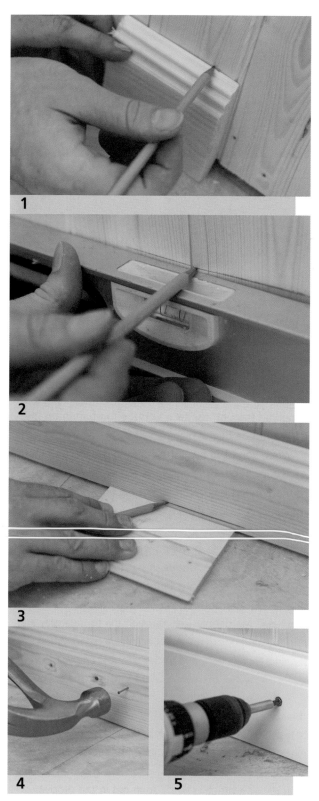

1

2

3

4

5

FIXING THE SKIRTING

If the floor is not level, measure down from the top batten in several places to find the lowest point of the floor. This will be the starting point for fixing the skirting board, and will allow you to cut (scribe) the underside of the skirting to fit any higher points in the floor around the rest of the room.

Offer up a skirting offcut to the panelling at the lowest point of the floor and mark the panelling along its top **1**. Continue this mark around the room, using the spirit level to make sure it is level **2**. This mark is for the top of the skirting. If the floor is uneven, the skirting will have to be scribed (shaped from the bottom) to match the pencil line **3**. Where the skirting meets at a corner you will need to mitre it.

Lightweight skirting can be fixed to the panelling using long pins **4**, which should go through the panelling into the lower batten. Fix heavy skirting with either losthead nails or countersunk screws **5**.

FINISHING THE TOP

There are various ways to finish the top of the panelling, but the simplest is to screw or pin and glue on L-shaped capping **6**. I prefer to buy some nice moulding or dado rail and glue and screw 50 x 25mm (2 x 1in) batten behind the top edge. I make up enough to finish the whole room, ready to glue and pin it to cover the top batten of the panelling. You need to cut the edges of the panelling, but this method allows all the preparation work to be done on a workbench (gluing, screwing and sanding) and when you mitre and scribe for the corners, the batten and dado will cut as one section.

6

FITTING A READY-MADE BATH PANEL

Position the panels vertically, using a spirit level or plumbline **1**, then mark a line on the floor along the length of both the panels **2**. The end panel fits behind the front panel at the corner joint. Remove the panels, and mark a second line 3mm (⅛in) inside the first. Screw battens to the floor along and inside this second line – a minimum of three battens for the front panel and two for the end one. For the batten to take the bottom flange, or turned-in edge, of the panels, you need to make a rebate. You can form this by placing a strip of packing the same depth as the flange beneath each batten. The packing will butt against the flange, while the batten above it will butt up to the inside of the panel itself. Reposition the panels, then drill 4mm (³⁄₁₆in) pilot holes through each one into the battens **3**. Secure the panels to the battens using decorative mirror screws **4**.

👍 TOP TIP If you have a router, cut a rebate in each batten to take the panel flange instead of using packing

tongue & groove panelling

Providing it's well planned and set out, tongue and groove panelling can look stunning in a bathroom, and is well within the capabilities of a DIY enthusiast. You can use it to re-create a period look, or to give a light, contemporary feel.

WHY PANELLING?

Panelling really comes into its own when you need to cover or conceal something ugly that otherwise would be difficult, and possibly expensive to correct. For example, some old tiling can be extremely difficult to remove. Pre-war tiling, especially, was often fixed with a strong sand-and-cement mix that normally requires some rather arduous club hammer and bolster work to get off. What's more, the render is likely to come off with the tiles, which means the wall will then have to be re-plastered before anything else can be done with it.

For this type of situation, panelling is a simple and attractive solution, providing you don't mind losing a few centimetres of room size. This is because the panelling is fixed to the wall by means of a batten framework.

I prefer the traditional tongue and groove pattern known as "bead and butt" to the more commonplace "V groove". For walls, I always use 12mm (½in) thick boards. There are also various MDF profiles to choose from, including one that imitates the traditional tongue and groove pattern that I like so much.

I find the best way to use panelling in a bathroom is to cover just over a third of the wall height. For example, in a bathroom with a 3m (10ft) high ceiling, the panelling should be about 1.2m (4ft) high from the floor.

The usual batten size is 50 x 25mm (2 x 1in), but you may have to increase batten size to 50 x 50mm (2 x 2in) if you need to conceal large pipework. Access panels may be required to reach isolating valves, stopcocks or waste pipe rodding eyes. Cutting around switches and sockets won't be a problem in the bathroom, because there shouldn't be any be in there. (If there are, they should be removed!)

👍 **TOP TIP For a bathroom, it makes sense either to buy pressure-treated battens and panelling, or to treat the timber yourself with a preservative before fixing it. This will give the timber some protection in an environment that is often moist and steamy.**

SETTING OUT AND FIXING THE BATTENS

Panelling that is less than 1.5 m (5ft) high will need just three fixing battens; one at the top, one at the bottom, and one exactly between these. Decide on the height of the panelling, then, using a spirit level, mark a level line around the room **1**; rotate the spirit level each time you mark the wall. This line marks the top of your panelling, so you fix the first batten immediately below it. Fix the bottom batten to the wall above the old skirting board in

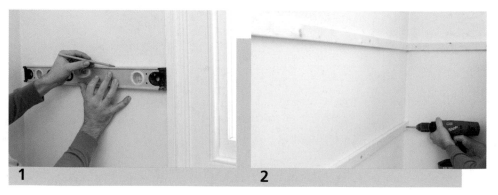

1

2

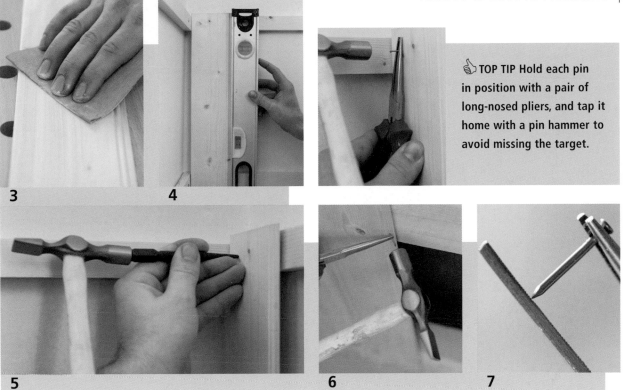

3   4

TOP TIP Hold each pin in position with a pair of long-nosed pliers, and tap it home with a pin hammer to avoid missing the target.

5   6   7

the same way; if there's isn't any skirting, fix the batten 50mm (2in) above the floor. Fix the middle batten an equal distance between the top and bottom battens **2**. Fix the screws 50mm (2in) from each end and at 400mm (16in) centres. If you wish to remove the old skirting, use a hammer to force a bolster chisel between the wall and the skirting, then insert a crowbar in the gap and lever off the skirting.

TOP TIP Before drilling the holes for the battens, mark the fixing positions on the wall, then check for pipes or cables using a detector.

FIXING THE PANELLING

Once the battens are in place, you can start on the panelling. Cut the first piece to length, and offer it up all around the room to check for size; any slight dips in the floor are unimportant, as the skirting will cover them. Cut roughly half the boards to size, and sand them down thoroughly with fine sandpaper on your bench **3**. Even though the timber is bought already planed, sanding down before fixing is crucial for a good finish. Starting at a left-hand corner, fix the first board in

position, with the groove facing inwards. Hammer a panel pin through the face into the top batten on the left-hand side of the plank.

When you have fixed the board to the top batten, use a spirit level to ensure it is vertical (plumb) **4**, then fix it to the two battens below. Use a pin punch to hammer the pins just below the surface **5**. To fix the remaining boards, "secret nail" them by pinning them through the tongue at an angle **6**. When you slip the groove of the next board over the tongue, it will hide the pins. To avoid splitting the tongue, first blunt the pin using the hammer or a file **7**. Alternatively, pilot-drill first, but ensure you don't drill too deeply – go through the tongue only and not the batten, or you won't achieve a fixing. After pinning each board, punch the pins just below the surface to allow the next board to fit over the tongue.

As you work, check regularly with the spirit level that the boards are vertical. Your guide for a horizontal level is to align the top of each tongue and groove board with the top batten. The tongue and groove housing will allow slight adjustments to be made if necessary. Don't worry – with practice you'll soon get in the groove!

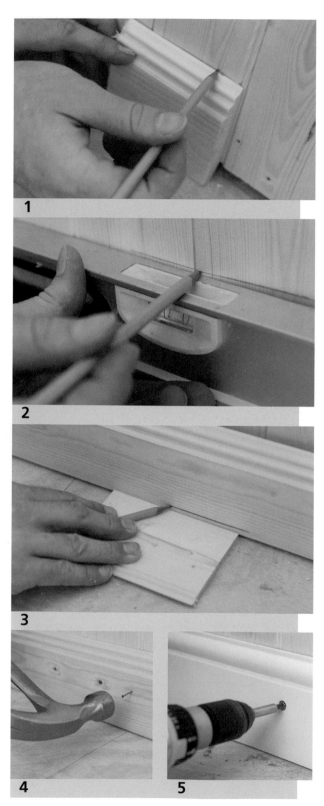

1

2

3

4   **5**

FIXING THE SKIRTING

If the floor is not level, measure down from the top batten in several places to find the lowest point of the floor. This will be the starting point for fixing the skirting board, and will allow you to cut (scribe) the underside of the skirting to fit any higher points in the floor around the rest of the room.

Offer up a skirting offcut to the panelling at the lowest point of the floor and mark the panelling along its top **1**. Continue this mark around the room, using the spirit level to make sure it is level **2**. This mark is for the top of the skirting. If the floor is uneven, the skirting will have to be scribed (shaped from the bottom) to match the pencil line **3**. Where the skirting meets at a corner you will need to mitre it.

Lightweight skirting can be fixed to the panelling using long pins **4**, which should go through the panelling into the lower batten. Fix heavy skirting with either losthead nails or countersunk screws **5**.

FINISHING THE TOP

There are various ways to finish the top of the panelling, but the simplest is to screw or pin and glue on L-shaped capping **6**. I prefer to buy some nice moulding or dado rail and glue and screw 50 x 25mm (2 x 1in) batten behind the top edge. I make up enough to finish the whole room, ready to glue and pin it to cover the top batten of the panelling. You need to cut the edges of the panelling, but this method allows all the preparation work to be done on a workbench (gluing, screwing and sanding) and when you mitre and scribe for the corners, the batten and dado will cut as one section.

6

CONSTRUCTING A TONGUE AND GROOVE BATH PANEL

1

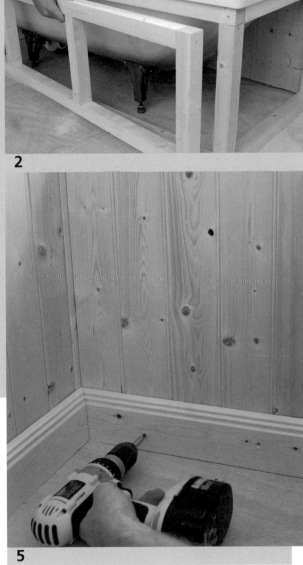

2

3

4

5

6

You can make a bath panel to match the tongue and groove panelling around the bathroom. First, fix a frame of 50 x 50mm (2 x 2in) timber to the floor and walls **1** (see page 93). Position the frame under the lip of the bath in order to protect any exposed cut edges from water. Make a second frame, strengthened with two cross members to fit within the first frame **2**. Cut the tongue and groove boards to cover the outer frame, but fix them to the inner frame by secret nailing them (see page 97). Made

this way, the boards attached to the inner frame form a removable panel that is fixed to the outer frame with three pairs of mirror screws all along the frame **3**. The whole panel can then be removed should there be any need to access the plumbing. Glue and pin a capping to the boards **4**, and cut and fix the skirting at the bottom **5**, mitring it into the corner. Finally, apply a mastic seam all around the panel to complete the job **6**. If you have to remove the panel at any time, re-apply the mastic seam.

PAINTING OR STAINING THE PANELLING

A quick and effective way to colour panelling is to use a preservative-and-stain mixture **1**. There are lots of water or spirit-based makes and colours on the market. I prefer to use a spirit-based preservative, but some water-based products are now considerably improved.

All you have to do is stir and apply. Two coats is the norm, but you can achieve a washed-out effect by rubbing all over with a cloth dipped in white spirit before the stain dries. The great thing about this finish is that any damage can be easily rectified with a little touching up. When it is time to redecorate, you simply apply another coat.

If you prefer a conventional paint finish, eggshell is probably the best choice for a bathroom, as it works and looks better than gloss in a moist environment. All new timber should have knotting fluid applied to any knots to seal them **2** before you paint. Once the knotting fluid is dry, apply one or two coats (as required) of primer **3**, then an undercoat, and finally a topcoat **4**. Allow each coat to dry thoroughly and rub it down very lightly with flour paper (very fine sandpaper), to remove any dust or bits, before applying the next coat. This may seem like extra work, but you'll end up with a lovely silky finish that makes all the difference.

2

3

**I suggest an
eggshell finish
for the bathroom**

1

4

concealing pipework

Surface pipework never looks attractive, so plan for any pipe runs above floor level to be hidden behind a "boxing" made from timber and plywood or MDF. The boxing can be decorated to match the rest of the bathroom, or topped with wood or stone to make an attractive shelf feature for displaying ornaments, fancy toiletries or floral displays. Remember to make access provision for any future maintenance requirements.

BOXING-IN PIPES

Try to keep pipe runs as short as possible and in a corner. For horizontal boxing, draw parallel lines on the wall and on the floor where you want the boxing to be, using a spirit level **1**. Screw 50 x 50mm (2 x 2in) or 50 x 25mm (2 x 1in) battens to the wall and floor at 600mm (24in) centres **2**, using plugs if fixing to a solid wall. If you need to box-in vertical pipes in a corner, simply run a length of batten vertically along each wall.

For the top and side of the boxing, cut two lengths from a 12mm (½in) thick sheet of MDF or plywood **3**. I find pre-finished water-resistant MDF is ideal; it is available in 2400 x 1200mm (8 x 4ft) and 3000 x 1200mm (10 x 4ft) sheets. Glue one of the MDF pieces to a third length of batten **4**, then screw them together with countersunk screws. Glue and screw the second MDF piece to the batten to form an L-shaped boxing.

👍 **TOP TIP Clamp the glued MDF and batten in position to enable accurate fixing whilst screwing 5.**

Position the L-shaped boxing over the fixed wall batten and scribe to match the wall if necessary. Fix the boxing in position with countersunk mirror screws, or use magnetic catches in case you need full access to the pipes at any time. If you are using mirror screws, drill pilot holes through the top of the boxing into the wall batten **6**. If necessary, cut an end piece from MDF and screw it in place to seal the end of the boxing. When boxing in horizontal pipes it may be necessary to cut out a space to allow for vertical pipes **7**. Decorate or tile the boxing to match the rest of the bathroom; the idea is to make the boxing as inconspicuous as possible.

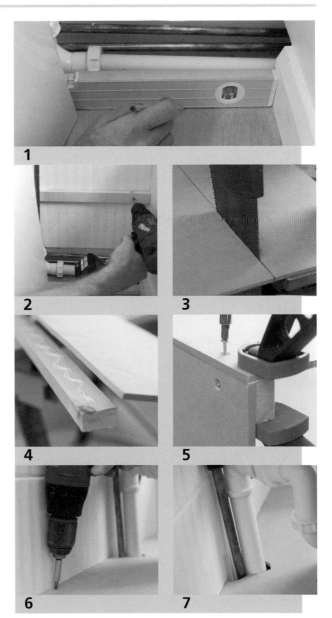

tiling bathroom wall surfaces

Calling in a tiling contractor can be expensive, and you may have to wait, so doing it yourself could save you money and get the job done quicker. If you follow some simple guidelines, there's no reason why you shouldn't be able to create a professional-looking job. Shop around for your tiles, so that you can buy some good quality tiling tools with the cash you save.

TILING SPLASHBACKS

A splashback is a small area of protective tiling, usually four or five courses, immediately above the washbasin or bath. If you want to create a splashback rather than tile a much larger area, tile up from the top edge of the basin or bath. Place spacers below the first course of tiles, and leave this joint ungrouted. When you have finished grouting everything else, you can fill it with a flexible silicone mastic sealant to make a waterproof joint.

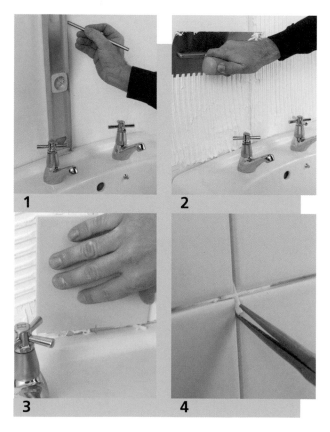

1   **2**

3   **4**

TOMMY'S GROUTING TIPS

Add the grouting powder to the water and mix until you have the consistency of double cream **A**, and apply with a rubber grouting float **B**. Remove excess grout before it sets, and polish up the tiles with a soft dry cloth, rubbing the joints over with a jointing tool **C** to smarten up the finish.

A

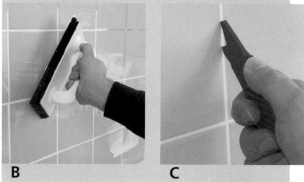

B   **C**

Mark the centre line **1**, then apply the adhesive. Use the straight edge of the trowel, working from the centre line out. Turn the trowel around and use the notched side to create vertical or horizontal lines in the adhesive, to aid adhesion **2**. As you fix each tile , give it a bit of a twist to ensure full contact **3** before straightening it, and insert spacers. You can push the spacers well below the tile surface, so they will be covered by the grout. Alternatively, you can push them in less deeply, and remove them before the adhesive is fully cured using a pair of long-nosed pliers **4**. If the splashback extends beyond the edge of the basin

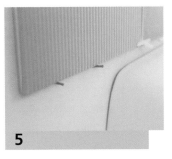

or bath, you may need to fix some pins or a temporary batten to hold them in position until the adhesive dries **5**. Clean off any adhesive on the tile surface using a damp sponge and leave overnight before grouting (see opposite).

Finally, apply silicone mastic sealant along the top edge of the basin or bath to create a waterproof seal **6**. Use your finger to smooth it off for a professional finish **7**.

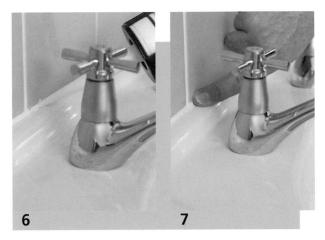

PREPARATION AND PLANNING

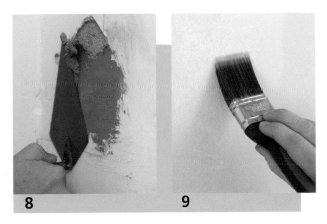

Prepare the walls before you start tiling. Remove any paper and scrape back flaky paint or plaster. A smooth painted or plaster surface will benefit from being sanded down with an electric sander, for better adhesion. You can make any plaster repairs using a two-coat system, which is undercoat (bonding) and finish plaster **8**. If the wall has to be replastered, it should ideally be with a fine sharp sand-and-cement render, because of the damp atmosphere, but don't apply the smooth finish coat where you are going to tile. A sand-and-cement plaster, with a wooden or plastic float finish, will provide the perfect surface for tiling.

👍 **TOP TIP Once an old wall surface has been prepared, apply a coat of diluted PVA adhesive to seal the surface and aid adhesion when tiling 9.**

HOW MANY TILES?

To work out how many tiles you need to buy, mark the outer edges of your tiled areas on the wall, then simply multiply the height by the width. Tiles are sold by the square metre, and the tubs of ready-mixed adhesive state clearly the area they can cover. It really couldn't be simpler, although it is always a good idea to buy a few extra tiles in case of breakages.

SETTING OUT

In order to determine where to start tiling, you need to measure each wall accurately. You can then make a gauge stick for setting out the tiles by marking a row of your chosen tiles onto a 50 x 25mm (2 x 1in) batten **10**,

1

including the spaces, or joints, between the tiles. You can decide the width for yourself, or use the width of a plastic tile spacer. Find the centre point of your wall and draw a horizontal line from either side of this point using a spirit level **1**. Holding the gauge stick vertically and then horizontally from that line, mark the wall so you have an equal cut tile at each end of every row and column **2**.

Before fixing the tiles to the wall, take some time to carefully mark out the best starting position to give equal cuts at either end horizontally, with the least amount of waste. Bear in mind that a tile cut into a corner should be at least a half tile; if less, reposition the starting point. Ideally, the cut tiles should be equal at both ends, and top and bottom where possible. Be patient! Even though

you'll be chomping at the bit to start tiling at this point, the "setting out" stage is very important. Stand back occasionally to check that you're happy with the guidelines you've made on the wall, and that everything looks symmetrical.

Once the setting out stage is complete, fix a temporary batten to the wall as a guide, and a support, for the first row of full tiles **3**. Fix it with plugs and screws, to allow for easy removal later. Always start tiling from the bottom up, so that each row of tiles is supported by the one below. Apply the adhesive using the straight edge of the notched trowel to cover about 2 sq m (2 sq yd) at a time, working from the centre line out. Then use the notched blade of the trowel to create horizontal lines in the adhesive, to aid adhesion **4**, and start fixing the tiles.

Give each tile a little twist as you fix it to ensure full contact with the adhesive **5**. As you work, fit spacers between the tiles to create uniform joints **6**, and continue tiling about 2 sq m (2 sq yd) at a time until the area has been covered. Clean off any excess adhesive from the tile surfaces with a wet sponge, and leave overnight to set before cutting and fixing the margin tiles (see opposite). Leave for another night before grouting **7**. Start grouting from a corner, working the material into the joints. Complete one wall at a time before moving on to the next one.

2   **3**   **4**   **5**   **6**   **7**

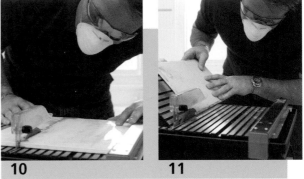

10   **11**

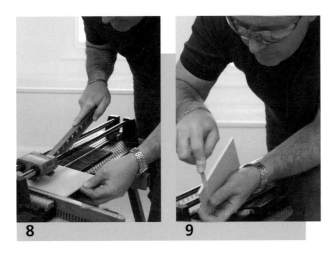

8   **9**

👍 TOP TIP I've heard it said that if you use a marker pen to mark tiles for cutting, it can bleed if the glazing is thin or cut. This has never happened to me, but to be on the safe side I use a wax crayon.

CUTTING TILES TO FIT

The main tiled area is called a field of tiles, and any cuts around the edges are called margin tiles. Straight cuts (margins) can be easily cut using a tile cutting jig. Run the cutting wheel along the marked line, then press the handle down to snap the tile **8**. The wheel needs only to score the glaze, so little pressure is required. Experiment with some scrap tiles to get the knack before cutting the proper margin tiles. Smooth down the cut edge of the tile using a small file. Rub in a downwards motion away from the face of the tile **9**, so avoiding breakout to the tile face. Always wear protective gear when cutting tiles, especially goggles to protect your eyes from flying splinters.

A wet tile cutter is very simple to use, and will cut both wall and sturdier floor tiles. The diamond-tipped cutting blade rotates through a reservoir of water, which cools the blade as it cuts **10**. This allows precision tile cutting, including mitring, to be carried out safely **11**, providing the cutter is of good quality.

CUTTING AROUND PIPES

I always try to plan pipe positions so that they will protrude through the tiles on or very close to a joint. This way, once the tile is marked, the cut can be easily made using a file saw. If you've done your preparations properly the pipes should emerge between the joints of two, or for the perfectionist, four tiles. If you're unlucky and the pipe emerges in the middle of a tile, don't worry. Transfer the pipe position onto the face of the tile using a set square, and mark around a pipe offcut. Cut the tile using a cordless drill and tile cutter attachment **12**. Apply adhesive to the back and slip over the pipe. The grouting will conceal the gap around the pipe.

👍 TOP TIP Vertical and horizontal corner joints to a bath, basin or shower should be filled with a silicone mastic sealant rather than grout. This will allow for expansion movement, whereas grout may crack and allow water to penetrate.

12

FIXING MOSAICS

Manufacturers have made fixing mosaics really easy. They are supplied in sheets, or panels, approximately 300mm (12in) square, with either a backing made from netting, or paper sheeting attached to the front. Mosaics with paper on the front are applied in the same way as those with netting (as shown here), but once the adhesive has set, soak the paper with warm water and a sponge, and lift the paper from one corner to remove it. Once all the paper has been removed you can grout as normal.

The setting out process for mosaics is the same as for normal tiling, that is marking out and working from the centre **1**. Apply the adhesive with a tiling trowel **2** then use the notched side to make adhesion lines **3** and fix the mosaic sheets **4**. You need to be accurate; mosaics with a backing sheet are difficult to move once in position, as the backing disintegrates after contact with the adhesive.

Ensure the sheets are flat, even and well attached to the adhesive by tapping them very lightly using a rubber mallet over an offcut of MDF or plywood **5**.

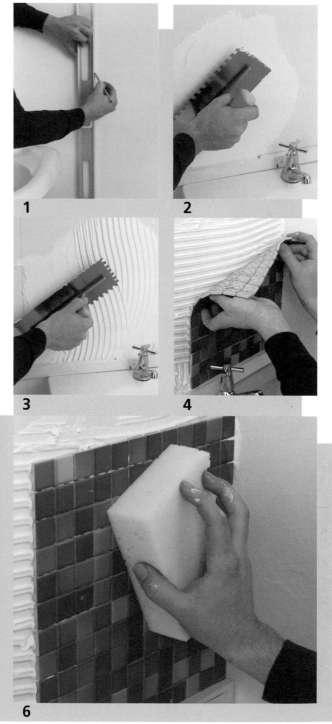

1

2

3

4

5

6

8

9

10

Clean off any excess adhesive before it sets with a damp sponge **6**. When you have fixed all the sheets, grout the mosaics in the same way as for normal tiles (see page 102) **7**. When the grout is dry, polish the tiles with a soft cloth.

To cut a mosaic sheet to size, use a craft knife to cut between the rows of tiles **8**. Make any shaped or awkward cuts with the help of a profile made from card, or even the protective paper **9**. Lay the profile over the mosaic sheet, mark around it with a marker **10**, then cut out the shape with a tile nibbler **11**. You can cut and fit the tiles individually, if necessary **12**.

11

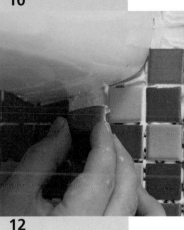

12

7

TOMMY'S ADVICE

You can create a stylish inlaid border by simply cutting mosaic sheets of a different colour or texture into strips, and fixing them all around the edge of the wall, a bit like a frame around a picture. This technique opens all sorts of possibilities: if you feel really creative, you could draw something on the wall, tile it with differently coloured mosaics then fill in around it to create a unique piece of art on your wall.

bathroom floor surfaces

You choice of bathroom floor should reflect three principal considerations: aesthetics, comfort and practicality. How easy it is to maintain and clean, and its safety in wet conditions are just as important as how it will look and feel underfoot when you pad around barefoot.

CHOOSING YOUR FLOORS

As well as making sure that the flooring is lovely to look at and comfortable underfoot, there are a couple of practical points to bear in mind. As there is a potential water danger to a wooden floor substructure, you need to ensure that any covering is water resistant and well sealed all around with a suitable silicone mastic sealant. Also, overlaid floors create a threshold difference at the doorway, which may cause people to trip, or tall people to bang their heads. (I should know, my head is like a moonscape with all the dents from door frames that I've bumped into over the years!) Make sure any overlaid floor doesn't create a trip hazard, by fitting a shaped threshold bar that marries the two levels. Finally, most bathroom floor coverings are simply laid on top of the existing floor structure, but ceramic and stone floor tiles require a plywood sub-floor to be laid first.

1

CERAMIC TILES

Ceramic tiles are a useful floor covering **1**. They are hard wearing, colourful, thin (so ideal for over-laying a floor) and impervious to water. Unless a form of under-floor heating is in place for the winter months, ceramic floors are cold underfoot, but this can be a boon in the summer.

NATURAL STONE

Natural stone flooring is definitely one of my all-time favourites. In the past, stone was expensive, due to its thickness and the cutting work involved, but today's machine-cut tiles are more practical and affordable. Modern producers have the technology to cut their products into thin tile form. Perfectly edged and pre-finished, these are suitable for any type of floor, and can be used with under-floor heating.

As a general rule, the harder and more impervious the stone, the easier it is to maintain, while the softer the stone, the more work is involved in keeping it looking good. Granite, certain types of marble, and slate are very hard-wearing and impervious to water. Sandstone, limestone and softer marbles have to be selected carefully following the supplier's advice. For example, limestone varies in both its durability and permeability, depending on how deep underground it was quarried. The tougher stuff comes from deep down and is less permeable than the prettier stuff, which comes from higher up and is less hard-wearing. All softer stone materials must be treated with two or three coats of water-proofing sealer. The most suitable stone floor finish is what is known as a honed finish, which is a sort of semi-flat finish that allows some grip underfoot.

VINYL TILES

Vinyl tiles provide an inexpensive easy-to-lay floor covering, and is far more forgiving than ceramic or stone, which means you can drop things on it! Available in 1sq m (1sq yd) packs, you simply pull off the backing paper to expose the adhesive and set the tiles in place.

SHEET VINYL

Because it is seamless (unless you have an enormous bathroom), sheet vinyl is ideal for areas where a lot of water may be splashed about. It is available in various densities, which provide different degrees of cushioning underfoot. Sheet vinyl is a quick flooring solution when laid by professionals, but the amateur will need to take great care to make a perfect job of it.

HARDWOOD FLOORING

Hardwood flooring is another favourite of mine, but it is expensive, and in a busy bathroom environment may require considerable maintenance. If you've set your heart on hardwood flooring, but it is out of your price range, consider an engineered hardwood floor. This is a slice of approximately 6mm (⅕in) thick hardwood bonded onto a substrate of MDF or plywood. It is available pre-finished but does allow for a limited amount of sanding down and re-polishing like traditional hardwood floors.

LAMINATE FLOORING

Laminate flooring is not the most suitable surface for a bathroom, even if it is bonded to a water-resistant sub-strate. Any surface water must be mopped dry immediately, to prevent it from seeping through the joints and causing the substrate base material to swell, which would then force the floor to lift up. Some people will no doubt opt for laminate in their bathroom, but personally I think the impracticality of the floor outweighs any benefit.

RUBBER SHEET AND TILES

Essentially for commercial use, rubber sheet flooring and tiles have been adapted for the domestic market, and now offer an enormous range of colours and textures **2**. Probably more suited to a contemporary bathroom style in a loft apartment rather than a period home, it is soft, warm, water-resistant, sound-absorbent and slip-resistant. It is also very easy to lay on a sub-floor with an adhesive.

CARPET

As far as I am concerned, this is a definite no-no, even the so-called "made-for-bathroom carpet". In my book, water and carpet just do not mix well, and unless real care is taken (even more so than with a laminate floor), you may have to rip it all out much sooner than you expect.

LINOLEUM

I have used lino for years and it is one of my favourite flooring materials. It is hard-wearing, comes in lovely colours and patterns, and is reasonably priced. It is also available impregnated with crushed aggregate, which produces great slip-resistance qualities. Lino is a sheet material with hessian backing; in a large bathroom it can be joint-welded to ensure total water resistance

CORK

Cork has some redeeming qualities. It's a great insulator of heat and sound, and it feels nice underfoot. Although it hasn't really moved on since the seventies, it is probably due for a timely re-appraisal.

2

PREPARING TIMBER FLOORS

Laying a soft floor covering directly over floorboards will eventually result in the floorboards showing through. This not only looks ugly but will cause the covering to wear very quickly. The same thing happens if you lay soft flooring over an uneven solid floor or one that is in poor condition.

To avoid this happening you need to create a smooth surface beneath the soft covering. You can cover a solid floor with self-levelling compound, but for floorboards you need to install a sub-floor comprised of sheet boarding. If you intend to lay hard flooring over floorboards, you'll need thicker board than for soft flooring, otherwise movement in the boards could flip the tiles off.

The downside to laying a sub-floor is the creation of a step at the room's threshold that could trip up anyone entering the bathroom. It may be necessary to shape a piece of wood to overcome this problem, to form a sort of mini-ramp (see page 88). Alternatively, you could buy a threshold bar from a specialist flooring shop.

FOR SOFT COVERINGS

Pliable floor coverings such as vinyl, cork etc, are best fitted over 6mm (¼in) plywood sheets that have been butted tightly together and fixed to the floorboards with 30mm (1¼in) ring nails.

👍 **TOP TIP Make sure that the combined thickness of the floorboards and plywood will take the ring nails without them emerging underneath and possibly hitting a water or heating pipe! Lift a section of floorboard, if necessary, to check.**

1

A slightly cheaper alternative to plywood sheets is to install hardboard sheets that have been dampened with water **1**. Fix them smooth side down with ring nails. Fixing damp sheets avoids the expansion that would result if they were fixed dry.

FOR HARD COVERINGS

If you want to lay ceramic, slate or granite tiles on a suspended wooden floor you must fit a 19mm (¾in) plywood sub-floor to stabilize the floor (see below). Check that the floor joists can carry the extra weight: they need to be at least 150 x 50mm (6 x 2in).

PREPARING FLOORBOARDS

First of all, check the existing floor for any loose or damaged floorboards **2**. If there any, this may be an indication that previous plumbing or electrical alterations have been carried out under the flooring. So, to avoid any potential disaster, it is vital to check whether there are any pipes or electric cables under the floorboards, prior to

2

fixing them. When fixing loose or replacement boards over which you intend to lay a sub-floor, it's safer and more accurate to first pilot-drill and then screw them down. Use a vacuum cleaner in all the nooks and crannies, then fix down any loose floorboards firmly.

FIXING A PLYWOOD SUB-FLOOR

The best method for fixing any bathroom sub-floor is to temporarily remove the toilet and washbasin before fixing. Alternatively, make a cardboard template of the fittings, transfer this to the plywood and simply cut out the shapes using a jigsaw or a handsaw. Use full-size 2400 x 1200mm (8 x 4ft) sheets of plywood where possible. Start by positioning a sheet along the longest wall, then mark the sheet in a grid fashion at 300mm (12in) spacings **3**. Pilot-drill and countersink holes at these points, so the screws are fixed level with or just below the plywood surface **4**.

Repeat this process until the floorboards are covered, staggering the sheets so that the joints do not align – this is called 'splitting the joints' **5**. If the skirting boards are

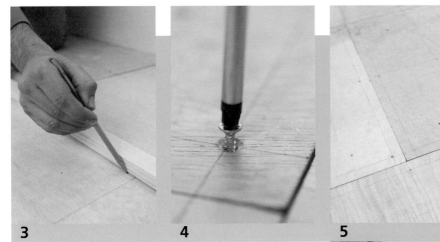

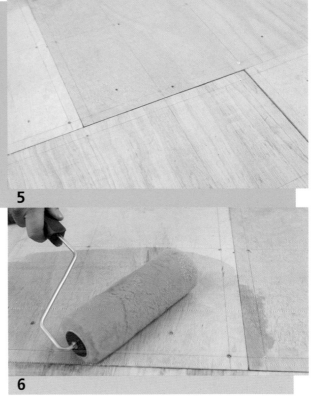

3 4 5

6

still in place, leave a 3mm (⅛in) gap all around the sub-floor as an expansion gap. If the skirtings have been removed, leave an expansion gap of 10mm (⅜in) all around between the plywood and walls; this gap will be completely concealed when the skirtings are re-positioned.

Finally, apply a coat of diluted PVA adhesive to the plywood sub-floor with a paintbrush or roller **6** before laying flooring tiles. This acts as a sealer and bonding agent for the adhesive.

LAYING CERAMIC TILES

If you don't feel confident about applying adhesive and fixing the tiles accurately, try the two-batten method – it's a fairly fool-proof way to lay a level tiled floor. First, you need to locate and mark the centre of the room on the floor, and the best way to do this is to use a chalk line. Measure and mark the halfway point on both sides of the room, then do the same for the end walls. Stretch and snap a chalk line between both sets of marks **7**. Where the lines cross marks the centre point of the room.

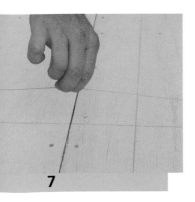

7

PREPARATION

Cut and fix a timber batten along the centre line to one half of the room. The batten must be level, so check with a spirit level and use thin packers where necessary. For accuracy and easy removal, I always plug and screw the batten down. Mark four tile widths, including the joints, onto another piece of batten and mark the floor at each end of the fixed batten. Using this gauge stick, attach a second batten to the floor, ensuring it is level with the first. This "racking back" or "pyramid" method sections the floor into roughly four quarters, working around to the final section with the door in it. I normally leave the margin tiles until the main floor has set. This allows you to correct any running off-line, no matter how slight, because each course left unchecked compounds any problem.

1 2 3

4

FITTING THE TILES

Apply adhesive to about 2sq m (2sq yd) of floor **1**, then score with a notched trowel. Position a corner of the first tile against the guide battens, gradually pressing the whole tile into position **2**. Lay the next tile abutting the first tile and the guide batten, then lay two more behind them in a square, using tile spacers **3** before laying the rest in a pyramid fashion. Lay the second quarter of the room in the same way. Lay all the full tiles before starting on the edges of the room. Using a straight piece of wood and a spirit level, check regularly that the tiles are all flush.

Release both battens and re-fix one of them to form the second bay for the last two quarters. The laid tiles act as one guide, the re-fixed batten as the other. Lay the tiles in the pyramid pattern as before. When you have laid all but the margin tiles, leave the floor to set overnight.

cutting margin tiles

When the adhesive has dried, cut in the margin tiles to go all around the edges. Cut the tiles using an electric wet saw, which can be hired or purchased relatively cheaply from DIY stores. It is easier to spread each tile separately with adhesive as you lay it. Allow the adhesive to dry and harden thoroughly for 24 hours before grouting the tiles.

grouting the floor

Release the tile spacers. Mix up the grout and press it between the tiles **4**, following the instructions on page 102. Clean off any excess immediately with a damp sponge.

👍 **TOP TIP To release the tile spacers quickly, use the flat end of a nail to hook them up.**

UNDERFLOOR HEATING

These days, underfloor heating is a viable DIY project. If you follow the steps shown here you can then lay the ceramic tiles as described. For timber floors you should prepare the surface as you would for tiling. The floorboards should be overlaid with tile-backer board (minimum thickness 10mm/⅜in) or plywood (minimum thickness 15mm/⅝in) screwed to the joists and noggins at no more than 300mm (12in) intervals. Tile-backer board is a specialized insulation board that will deflect rising heat to ensure the system works efficiently. When using plywood, first brush one coat of a specialist coating called BAL Bond SBR onto the surface and allow it to dry. Ensure that the underside of a timber floor is adequately ventilated.

installation of the system

To determine which size is required, measure the free floor area where the underfloor heating system is to be installed. Cut a groove using a router, approximately 10mm (⅜in) deep and wide, into the floor **5** and wall. This is for the floor sensor and tube. The tube allows easy removal of the floor sensor if required. Once the mat has been tested, lay

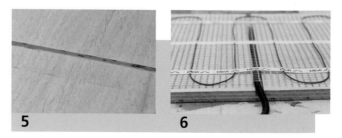

5 6

7

it on the floor in strips across the room **6**. At the end of a mat run, cut and turn the mat, laying the next piece beside the first. Do not cut the red heating cable. The Devimat system is self-adhesive and will stick to the floor providing the surface has been well prepared as instructed above. Cover with a flexible self-levelling compound, following the manufacturer's instructions and allow to dry **7**. Then tile using a rapid setting flexible adhesive and grout.

👍 **TOP TIP Electrical connections to the floor sensor, which is attached to the wall, should be made by a qualified electrician.**

LAYING VINYL TILES

Unpack the tiles from the boxes, stack them in the room where you intend to use them, and leave them for at least 24 hours to acclimatize before you lay them. This means the tiles will be flexible and easy to work with. Check whether the tiles have a pattern that follows a particular direction (some tiles are marked on the back with arrows as a guide). To check the pattern is correct, lay some tiles before fixing, without taking off the backing.

When you are ready to lay the tiles, snap a chalk line to make two bisecting lines across the centre of the room (see page 111). The first lot of tiles should be laid to the line, to cover the half of the room away from the door. I lay all the full tiles first, before the margins, to act as a guide for alignment.

To lay a vinyl tile, peel off the backing paper, then position a corner of the tile against the crossing guidelines and gradually press it into position **8**. Lay the next tile on the other side of the centre line, abutting the first tile and

the line **9**. Fix two more tiles immediately behind the first two, to form a square. Continue to lay tiles around this square in a pyramid fashion, until you have covered one half of the room, laying only full tiles and leaving the edges. Lay the second half of the room exactly as you did the first.

cutting around the edges

I think that it is fair to say that in all the years I have worked in the building industry, I cannot recall a single building that was perfectly square. This is why floors are laid from the middle to the edge. Fitting cut tiles around the edge of a newly laid floor is known as trimming margin tiles. To cut one, first lay a loose tile exactly on top of the last tile laid. Place a second tile on top but with its edge touching the wall. Draw a line with a wax crayon along the edge of this tile, to mark the tile below **10**. Cut along the line using a craft knife. Finally, fit the cut-off tile portion into the margin for the perfect fit.

8   9   10

ventilation

In the past, fresh air moved freely around our houses and moisture escaped naturally through cracks and open windows, but homes today are much more airtight than they used to be, making ventilation a priority. We need fresh air to breathe and if moisture, especially all that steam produced in our modern bathrooms, cannot escape it will condense and cause problems.

KEEPING THE AIR CIRCULATING

In earlier days, the whole house was built to breathe. Windows let in small amounts of fresh air when they were closed, and roof voids were vented through the slates and tiles before underfelt was used. However, the combined introduction of central heating, double glazing and insulation, particularly to old housing stock, has presented a whole new range of problems.

I'm not saying that this technology is a bad thing, but when it is installed, certain provisions must be made for efficient ventilation, particularly in rooms that produce lots of moisture. Without a constant exchange of air, the room becomes stuffy and the air moisture content increases, and eventually condenses, which may cause serious problems. Furthermore, adequate ventilation for a bathroom (or indeed any room) with a gas water heater is of the utmost importance; heater vents must never be covered or blocked, and they must be checked annually by a CORGI-registered fitter. Failure to comply could lead to a build up of fumes, which could cause serious injury or death by asphyxiation.

The simplest way to improve ventilation in the bathroom is to leave the bathroom door and window open at all times. If these solutions compromise your home security or family privacy, you could try installing a window vent instead. A particularly good solution would be to combine a brick air vent, let into an outside wall, with a trickle vent cut into the frame of a double-glazed window to allow the passage of fresh air (see opposite).

INSTALLING A BRICK AIR VENT

When installing an airbrick in a cavity wall, you must also fit a cavity liner between the outer and inner leaves of the wall. Otherwise the moist air will simply be vented into the cavity and cause damage elsewhere!

Ideally, a brick air vent should be fixed in the upper third of the wall, because that is where hot air collects, and away from a boiler vent. Check for buried pipes and cables using a detector, then mark the position for the vent outside. Using a hammer drill fitted with masonry bit, drill a hole through the centre of the marked brick into the bathroom **1**. Check that the vent's position inside the bathroom is correct. Outside again, drill a series of holes in the mortar around the brick to weaken it **2**. Then cut out

1   2   3

WINDOW VENTS

The three types of window vent provide quick-fix solutions to a ventilation problem. All are easy to fit, but require a hole to be cut in the glass. If you can't cut the glass yourself (you need a glass-cutter), call in a glazier to cut it on site, or remove the glass from the frame and take to the shop.

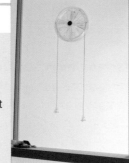

• A permanent window vent is cheap and easy to fit, but draughty.
• A wind-driven window vent (shown here) is cheap, easy to fit and costs nothing to run, but it can be noisy, sometimes developing an irritating squeak.
• An electricity-powered window vent is easy to fit and is the most efficient type. It requires power, however, so there are running and maintenance costs to consider.

hammer and bolster chisel (remember to wear safety gloves and glasses) to carefully cut out the brick **3**.

Clean all debris from the cavity **4** then splash water inside it, using a paintbrush, and leave it to soak in for 10 minutes. Trowel a bed of mortar (5:1 soft sand to cement) on the sides and base of the cavity **5**, and then onto the top of the air brick and carefully position this flush with the brickwork **6**. Tidy up the mortar joint to match the surrounding brickwork: for a weathered joint, use a pointing trowel; for a recessed joint, rub a jointer, a bent piece of copper pipe, or even a piece of hose over it.

Inside, fit a louvre grille over the opening, preferably one with an integral fly screen. Screw the fly screen in place first **7** then clip on the grille **8**. It is a good idea to apply adhesive as well as use screws to ensure an insect- and dust-proof closure. Wipe off any excess adhesive and leave to dry. You can paint the grille to match the wall.

👍 **TOP TIP Brick air vents are available in various colours and in three sizes: 225 x 75mm (9 x 3in), 225 x 150mm (9 x 6in) and 225 x 225mm (9 x 9in). Select a shade to match your brickwork as closely as possible.**

the inner skin. If the wall is tiled inside, keep the removed tiles for cutting down to size and refitting once the hole has been cut through. Outside once again, use a club

A TRICKLE VENT

Double-glazed windows fitted in a room with a gas water heater must have a combustion air vent in the head of the frame to ensure a constant supply of fresh air. To fit such a vent, known as a trickle vent, in an existing window, cut a narrow slot in the frame or sash and cover it, inside and outside, with a two-part plastic vent fitting.

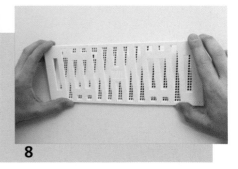

4

5

6

7

8

1

always switch off the power supply before starting any electrical work

2

3

4

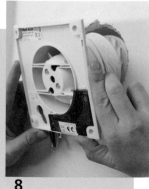

5

6

7

8

ELECTRIC EXTRACTOR FANS

If your bathroom is particularly prone to condensation, it may be worth installing an electric extractor fan to expel moisture-laden air quickly, and without draughts.

A window or outside wall is required for the fan, and the position is quite critical. To remove condensation, fresh air must circulate around the whole room, so ideally the fan should be positioned at a high level, in the corner opposite the door. It needs to be high up to remove hot, moisture-laden air as it rises. If the room contains a gas water heater without a balanced flue, check with a CORGI-registered gas fitter that there is enough replacement fresh air. Otherwise, when the fan is switched on, the fumes from the appliance may be drawn into the room, rather than exit through the flue. A balanced flue appliance takes its air supply directly from outside.

Most electric fans have an integral switch; alternatively, a switched fused connection unit can be wired into the supply during installation. The power supply can be provided by a spur taken from a nearby ring main, and a qualified electrician can do this for you. Some fans have a built-in speed control to regulate extraction speeds, and a timer that allows the fan to switch off after a suitable interval. Other fans are controlled by the level of humidity in the room, and operate accordingly.

Axial fans are ideal for fitting in a window, or with ducting through a wall, but if the ducting needs to be long, a centrifugal fan will be required. Select a fan with a

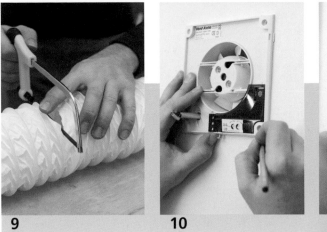

9　　**10**　　**11**　　**12**

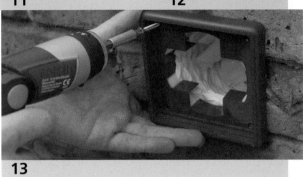

13

set of external shutters to prevent draughts when the fan is not in use. Low-voltage fans with their own transformers are for fitting directly over a shower, which complies with safety regulations; these extract moisture via a centrifugal fan through ducting to the outside.

SIZE OF FAN NEEDED

To work out the fan capacity required, multiply the volume of the room (length x width x height) by the number of air changes required per hour. An average bathroom requires 6–8 air changes, but this almost trebles to 15–20 changes when a shower is installed. A toilet requires 6–10 changes.

FITTING A WALL-MOUNTED FAN

Use the ducting to mark the position on the bathroom wall **1**. Check for buried pipes and wires with a detector. Make a centre mark, then using a hammer drill fitted with a long masonry bit, drill through to the outside **2**; keep the drill level. Centre the ducting over the hole on the outside and mark the wall **3**. Drill a series of holes to weaken the brickwork and cut out using a club hammer and cold chisel **4**. (Drilling and cutting from the outside in, reduces the amount of damage.) When approximately halfway through, or when reaching the cavity in a double wall **5**, go inside and repeat the process **6**. Remove the inner section, causing the minimum plaster damage.

All fans are generally fitted in the same way. Attach the self-adhesive sealing strip to the spigot end of the duct backplate **7**. Insert the ducting in the cut hole, with the backplate against the wall **8**. Mark the length of ducting, making allowance for the outer grille spigot, and cut it

using a hacksaw **9**. Fit the ducting in the hole. Mark the fixing holes on the wall **10**, then drill and plug these. Feed the electricity cable through the backplate before fixing it to the wall **11**. Wire up the fan according to the manufacturer's instructions. Mark the fixing positions on the second spigot end. Before fixing, squeeze some gap-fill expanding foam around the ducting to seal any voids **12**. Screw to the wall **13**. Seal any gaps outside with mastic. Screw the grille in place inside **14**, and fix the plate outside, covering it with the external shutters **15**.

14　　　　**15**

final bathroom finishes

When you've spent all your hard-earned cash on improving your bathroom, or even re-fitting it completely, it needs a really good decorating job to finish it off properly. If you don't want to pay out for a professional to do it for you, take the time and care to do a first class job yourself. The right paints, good brushes and careful preparation are the key to good paintwork.

PAINTING AND DECORATING

Oil-based paints like gloss and eggshell are suitable for bathrooms, as they seal the walls and ceiling from moisture. Unfortunately, water vapour tends to condense on their shiny surfaces and run down the walls and paintwork, as it does on ceramic tiles. To combat this, paint manufacturers have produced paints with anti-fungal and anti-condensation properties specifically for use in bathrooms. Though it has the same appearance, unlike matt emulsion bathroom paint is a tough finish and does allow for some cleaning of marks from the surface. It is available in a range of colours.

FIRST JOB

Carefully cover the bathroom with dust sheets, then prepare the ceilings and walls. Remove any flaking and loose paint with a scraper **1**, and fill any screw holes or small undulations with a proprietary filler. Any major plaster damage will have to be re-plastered. Sand down the walls with medium then fine sandpaper, either by hand

the quality of the finish depends entirely on the preparation

using a sanding block **2**, or with an electric sander. Once the surface is perfectly smooth, apply a coat of diluted PVA adhesive to seal the surface. Thoroughly clean the bathroom of all dust: remove the dust sheets and shake them outside, mop over the floor, and sponge down any surfaces to remove all traces of dust. Replace the dust sheets, then open the windows and door to keep the room well ventilated whilst you are painting. Paint the ceiling first, applying two coats for a professional finish, then coat the walls. The walls should always be cut into the ceiling, rather than the other way round **3**. Again, apply at least two coats for the best finish.

PAINTING WOODWORK

Painting woodwork in the bathroom is pretty much the same as anywhere else in the home, with the exception that any newly installed wood should be treated on the back as well as the front with either primer or preservative.

I know I'm repeating myself, but the quality of the finish depends entirely on the preparation, so spend an extra day or two on the preparation before applying any paint.

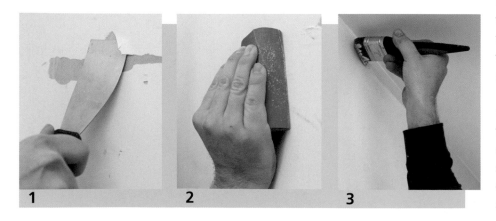

1   2   3

4

> 👍 TOP TIP An electric palm sander is a very useful piece of equipment (left). As its name suggests, it's a small sander that fits into the palm of one hand. It is extremely simple to operate, and its size makes it easy to get into awkward areas. It's especially useful for sanding down fixed woodwork.

Woodwork should be thoroughly sanded down, using a medium-grade sandpaper followed by a fine-grade one to achieve a smooth, professional finish. If you have filled any wood, leave this for 24 hours before sanding.

You'll also need to seal any visible knots with knotting fluid (see page 100). When this has dried you can start painting. Apply decorating mastic to gaps or around any joists, removing any excess with a damp sponge. Now apply a good coat of primer, and when it has dried, lightly rub it over with flour paper (very fine sandpaper) to remove any stuck on brush hairs, dust or dirt. Then apply two coats of undercoat followed by one topcoat, lightly rubbing over each undercoat once it has dried with flour paper. Brush along the grain **4**, rather than across it.

WINDOW TREATMENTS

For obvious reasons, a bathroom window needs to address the problem of privacy. Where there is no frosted pane already in place, you'll have to do something about it. Net curtains are largely impractical – any fabric is liable to go very limp in a bathroom's steamy conditions. You could replace a clear pane with frosted glass, at least in the lower section of the window. Alternatively, you could stick frosted plastic film over it, which would save you the effort and money involved in replacing the glass.

BLINDS

Another option is to put up a blind. Available in a large selection of colours, sizes and styles, blinds are normally cheaper to buy than curtains. They are also easier to install and tend to be made from humidity-tolerant materials. You could try a simple roller **5** or a roman blind. Venetian blinds are notorious as dust traps, but they do have the advantage that you can change from privacy to a view at the flick of a wrist. For a bathroom, a plasticized type would be preferable to a wooden one.

5

LIGHTING

The lighting in a bathroom plays an important role, and the wrong type can make it stark and soulless. Choose fittings to complement your bathroom's overall style.

Steam in a bathroom can cause bulbs to shatter, so from a safety aspect, all bulbs must be housed in moisture-proof, enclosed light fittings. Many bathroom cabinets now have an integral light and shaver socket **1**. Light switches within the bathroom must be ceiling-mounted pull-cords; a wall-mounted switch may only be positioned outside the bathroom. (See pages 22–24 for more details about using electricity in the bathroom.)

TYPES OF FITTINGS

Safety regulations (see page 23) severely restrict the choice of bathroom light fittings to basically three types:
• A self-contained central light fitting. Like the traditional glass or plastic globe which screws into a base plate, this

1

protects the bulb and holder.
• Mains-voltage recessed spotlight. These work off the existing lighting circuit.
• Recessed spotlights (down lighters are the common name). These low-voltage lights are great for general lighting. They are powered from a 12-volt transformer that may be linked to the main power circuit by a spur with a 5-amp fused connection unit. Alternatively, and preferably, the lights and transformer can both be powered from the existing lighting circuit.

👍 **TOP TIP Any bulbs should be vapour proof to avoid any possible blow outs.**

BATHROOM ACCESSORIES

Accessorizing is a very important part of bathroom design, and it's something that can be done easily to transform a fairly unexciting but functional bathroom into something to wonder at. Where you position accessories can make a world of difference, and it's surprising how many people put their fixed accessories in the wrong place entirely. The toilet roll holder, for example should be positioned at a comfortable height near the pan, for obvious reasons, but not so that it causes an obstruction; it may look really nice on the opposite wall, but that's not very practical. A fixed soap dish should be fixed directly over the hand basin, to the right if you are right handed, the left if left handed; any water draining through the dish can then drip onto the hand basin rather then the floor. Towel rails should be positioned close to the bath or shower, while a smaller rail next to the hand basin can hold the hand towel **2**. Ideally, the bath towel rail should be fixed above a radiator, or you could fit a towel rail radiator instead (see pages 34–35).

2

3

In my shower, I have hung up a three-layered chrome basket to hold all the different shampoos and conditioners everybody uses. It also holds various sponges and a back washer, all at high level, away from the shower spray.

Mirror-fronted bathroom cabinets have been around for years, but they have changed a great deal and are still very useful, especially positioned above the hand basin **3**. They provide perfect storage for much-needed items that are better kept unseen. Buy a solid, well-made cabinet that will stand up to a lot of use, and choose one with a built-in light and shaving socket, so that it can charge the razor and electric toothbrush out of sight (see left).

FITTING A GLASS SHELF

Permanent accessories like a shelf need to be firmly fixed to the wall due to the weight placed on them and the amount of regular use they will get. Offer the shelf up to the wall, mark its position, and check it is level **4**. If you are fixing through ceramic tiles, stick masking tape over them, then mark the fixing positions. The tape will help to stop the drill sliding off course and scratching the tile. Fit a masonry bit and, with the drill in the drill (not hammer) position, drill through the masking tape and the tile, then change to the hammer setting to drill into the masonry.

4

👍 **TOP TIP For a clean, no mess job, use masking tape to attach an open envelope under the fixing positions to catch falling masonry dust.**

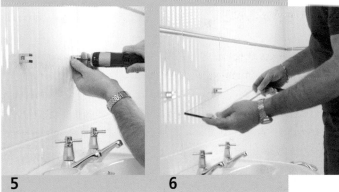

5   **6**

👍 **TOP TIP Match the drill size to the plug size; red plugs are normally drill size 6, brown plugs drill size 7.**

Remove the masking tape and insert the plugs provided with the shelf. When fixing to plasterboard, or into a void, use the correct type of anchor to get a firm fixing. Screw the shelf holders home **5**. Don't over tighten the screws or they can snap; once they are tight, an extra half turn is all that's needed. Take your time to avoid slipping with the screwdriver and scratching the tile. Position the shelf **6**.

glossary

Airlock A blockage in a pipe caused by a trapped air bubble.

Appliance Any machine or device that is powered by electricity.

Back-siphoning Siphoning of part of a plumbing system due to the mains water pressure failing.

Banjo unit A combined waste and overflow unit.

Base coat A flat coat of paint over which a layer of glaze is applied.

Batten A narrow strip of wood, usually fixed to a wall to act as a support for a unit or shelving.

Bevel Any angle at which two pieces of wood meet, other than a right angle.

Bore The hollow part of a pipe.

Butt joint A simple joint where two pieces of wood are fixed together with no interlocking parts cut in them.

Cam and stud fixing A simple fixing used in flat-pack construction.

Cap-nut A nut used to tighten a fitting onto pipework.

Cavity wall A wall made of two separate, parallel masonry skins with an air space between.

Chamfer A flat, narrow surface along the edge of a workpiece, usually at a 45° angle to any adjacent surfaces.

Chase A groove cut in masonry or plaster for electrical cabling or pipework.

Circuit A complete path through which an electric current flows.

Conductor A component, usually a length of wire, along which an electric current will pass.

Consumer unit The box containing all the fuse ways that protect the individual circuits in the house. The main on-off switch is located here, enabling you to isolate the power supply to the whole house.

Counterbore A tapered recess that allows the head of a screw or bolt to lie below a surface; also to cut such a recess.

Countersink To cut a tapered recess that allows the head of a screw or bolt to lie flush with a surface.

Cup To bend as a result of shrinkage; usually referred to as across the length of a piece of wood.

Damp-proof course (DPC) A layer of impervious material that prevents moisture rising through a floor or in a wall.

Earth A connection between the earth or ground and an electrical circuit; also a terminal to which this connection is made.

Extension lead A length of electrical flex for the temporary connection of an appliance to a wall socket.

Face edge A woodworking term for a surface that is planed square to the face side (see below).

Face side A woodworking term for a flat, planed surface from which other angles and dimensions are measured and worked.

Fence An adjustable guide to keep the cutting edge of a tool a set distance from the edge of a workpiece.

Flat-pack Furniture or units supplied in pieces and assembled by the purchaser, using knock-down fittings.

Four-way A block of four electrical sockets connected to a wall socket by an extension lead.

Free-standing Furniture or units that are not built-in or fixed to a wall or floor.

Fuse board A unit where a main electrical service cable is connected to the circuits in a house; also a term covering a meter, consumer unit, etc.

Galvanized Covered with a protective coating of zinc.

Grain The direction of wood fibres in a particular workpiece; also a pattern on the surface of timber made by cutting through the fibres.

Groove A long, narrow channel cut in plaster or wood; in the latter, this follows the direction of the grain.

Grounds Strips of wood fixed to a wall to provide nail-fixing points for skirting boards, etc.

Housing A long, narrow channel cut across the general direction of wood grain to form part of a joint.

Insulation Material used to reduce the transmission of heat or sound; also a non-conductive material around electrical wires or connections to prevent the passage of electricity.

Isolating valve A valve used to shut off water from a particular room or appliance, so as not to have to turn off the entire water system.

Joist A horizontal wooden or metal beam (such as a RSJ) used to support a structure such as a floor, ceiling or wall.

Key To roughen a surface to provide a better grip when it is being glued; also the surface so roughened.

Knock-down (KD) Another name for flat-pack furniture or units.

Knock-down (KD) fittings Fittings and fixings supplied with flat-pack furniture or units by the manufacturers, typically including screws, bolts and cam and stud fixings.

Knotting Sealer, made from shellac, that prevents wood resin bleeding through a surface finish.

Knurled On a knob or handle, a series of fine grooves impressed into an edge or surface to improve the grip when turned or handled.

Laminate Two or more sheets of material bonded together; or the top waterproof sheet of the bonded sheets used as a work surface; also to fix such sheets together.

Lintel A horizontal beam used to support the wall over a door or window opening.

Lipping A decorative strip applied to the side edges of laminated boards.

MDF Medium-density fibreboard, a man-made sheet material that can be worked like wood and is used as a substitute for it.

Mitre A joint between two pieces of wood formed by cutting 45° bevels at the end of each piece; also to cut such a joint.

Noggin Horizontal reinforcing timber fixed between the vertical studs in a stud partition wall.

Pilot hole A small-diameter hole drilled to act as a guide for a screw thread.

Primer A coat of paint applied to wood or metal to seal it and act as a first coat.

Profile The outline or contour of an object.

PTFE tape Tape made from polytetrafluorethylene, used to seal threaded plumbing fittings.

RCD Residual circuit device, a device that monitors the flow of electrical current through the live and neutral wires of a circuit.

Rebate A stepped rectangular recess along the edge of a workpiece, usually forming part of a joint; also to cut such a recess.

Reveal The vertical side of an opening.

Rising main A pipe that supplies water under mains pressure, usually to a roof storage tank.

Score To scratch a line with a pointed tool.

Scribe To copy the profile of a surface on the edge of sheet material to be butted against it; also to mark a line with a pointed tool.

Sheathing An outer layer of insulation on an electrical cable or flex.

Short circuit Accidental re-routing of electricity to earth, which increases the flow of current and consequently blows a fuse.

Silicone mastic A non-setting compound used to seal joints.

Spur Branch cable that extends an existing electrical circuit.

Stud partition A timber frame interior dividing wall.

Template A cut-out pattern, made from paper, wood, metal etc, used to help shape a workpiece accurately.

Terminal A connection to which bared ends of electrical cable or flex are attached.

Trap A bent section of pipe below a bath, sink, etc., containing standing water to prevent the passage of gases.

U-bend A waste pipe, or part of one, shaped like a U, used as part of a trap.

Undercoat A layer or layers of paint used to cover primer and build up a protective layer before a top coat is applied.

index

suppliers

BATHROOM SUPPLIERS

AMBIANCE BAIN
bathroom suppliers
tel: 0870 902 1313
web: www.ambiancebain.com

AQUAPLUS
bathroom suppliers
tel: 0870 201 1915
email: info@aquaplussolutions.com
web: www.aquaplussolutions.com

BATHSTORE.COM
bathroom suppliers
tel: 0208 748 6656
email: hammersmith@bathstore.co.uk
web: www.bathstore.com

BATHWISE
bathroom suppliers
tel: 0208 840 9313
email: sales@bathwise.co.uk
web: www.bathwise.co.uk

DARYL SHOWERING
showers, shower enclosures and trays
tel: 0808 100 1627 customer enquiries
email: daryl@daryl-showers.co.uk
web: www.daryl-showers.co.uk

GROHE
taps and showers
tel: 0870 848 8877
web: www.grohe.co.uk

HANSGROHE
hansgrohe   taps and showers
tel: 0870 770 1972 for sales and stockists
email: sales@hansgrohe.co.uk
web: www.hansgrohe.co.uk

IKEA
fitted bathroom & accessories suppliers
tel: 0845 355 1141 for stockists
email: via website
web: www.ikea.co.uk for stock availability

JACUZZI UK
JACUZZI UK bathroom suppliers
Fordham, Niagara and Jacuzzi are all Jacuzzi
UK product brands
tel: 01274 654 700
email: pr@jacuzziuk.com
web: www.jacuzziuk.com

Fordham
tel: 01274 654 700
web: www.fordham.co.uk

Jacuzzi
tel: 01782 717 175
web: www.jacuzzi.co.uk

Niagara
tel: 01274 654 700
web: www.fordham.co.uk

KEUCO
bathroom suppliers
tel: 01442 865 220
web: www.keuco.de

LAUFEN
bathroom suppliers
tel: 01386 422 768 for general enquiries
web: www.laufen.co.uk

LEFROY BROOKS
bathroom suppliers
tel: 01992 708 316
web: www.lefroybrooks.co.uk

MANHATTAN SHOWERS
shower enclosures and trays
manhattan tel: 01282 605 000
web: www.manhattan showers.co.uk

THE METROPOLITAN SHOWER CO.
shower enclosures and trays
tel: 01282 606 070
web: www.metropolitanshowers.co.uk

SANIFLO
suppliers of toilet macerators
tel: 0208 842 0033
email: sales@saniflo.co.uk
web: www.saniflo.com

SCHNEIDER
bathroom accessories suppliers
tel: 0041 433 777 878
web: www.wschneider.com

TRITON
electric shower & pump suppliers
tel: 0800 0644 645 for brochures
web: www.tritonshowers.co.uk

TSUNAMI
bathroom suppliers
tel: 0207 408 2230
web: www.tsunami-interiors.co.uk

TWYFORD BATHROOMS
bathroom suppliers
tel: general info 01270 879777
email: sales@twyfordbathrooms.com
web: www.twyfordbathrooms.co.uk

VILLEROY & BOCH
bathroom suppliers
tel: 01625 525 202
web: www.villeroy-boch.com

FLOORING & TILES

AMTICO
flooring suppliers
tel: reader enquiries 0800 66 77 66
email: customer.services@amtico.co.uk
web: www.amtico.com

BAL
tile adhesive & grout suppliers
tel: 01782 591 160 for customer service
email: info@building-adhesives.com
web: www.building-adhesives.com

DALSOUPLE
rubber flooring suppliers
tel: 01278 727 733 for stockists
email: info@dalsouple.com
web: www.dalsouple.com

DEVI
devimat underfloor heating suppliers
tel: customer enquiries 01359 243 514
email: devimat@devi.co.uk
web: www.devi.co.uk

FORBO
flooring suppliers – Novolon, Marmoleum
and Cushionflor are all Forbo products
tel: customer enquiries 0800731 2369
email: via websites or info.uk@forbo.com
web: www.marmoleum.co.uk
web: www.cushionflor.co.uk

JOHNSON TILES
floor & wall tile suppliers
products available worldwide
tel: 01782 575 575 customer services
email: sales@johnson-tiles.com
web: www.johnson-tiles.com

(mosaic from Colchester Tiles can be
obtained from Bathstore)

OTHER SUPPLIERS

B&Q major DIY store for everything
tel: 0845 222 1000 for stockists
web: www.diy.com

MK ELECTRIC
electrical suppliers
tel: mk helpline 0870 240 3385
web: www.mkelectric.co.uk

MYSON
radiators and towel-rail suppliers
tel: customer enquiries 0191 491 7530
email: sales@myson.co.uk
web: www.myson.co.uk

VENTAXIA
ventilation equipment
tel: 01293 526 062 customer enquiries
email: info@vent-axia
web: www. vent-axia.com

CROWN
paint suppliers
tel: 0870 240 1127 paint talk helpline
web: www.crownpaint.co.uk

ATLAS COPCO (brand name Milwaukee)
power tool suppliers
tel: 01442 222378 for nearest dealer
email: milwaukee@uk.atlascopco.com
web: www.milwaukee_et.com

BLACK & DECKER
power tool suppliers
tel: 01753 511234 for customer helpline
email: info@blackanddecker.co.uk
web: www.blackanddecker.co.uk

DE WALT
power tool suppliers
tel: 0700 433 9258 for stockists
email: via website
web: www.dewalt.co.uk

MARSHALLTOWN & ESTWING TOOLS
power tool suppliers (supplied by Rollins)
email: sales@rollins.co.uk
web: www.rollins.co.uk

SCREWFIX DIRECT
tool suppliers
tel: 0500 414141
email: online@screwfix.com
web: www.screwfix.com

PLASPLUGS
tiling tool suppliers
tel: sales enquiries 01283 53 00 00
email: sales@plasplugs.com
web: www.plasplugs.com

photography credits

2–5 All David Murphy; **7–12** All Mike Newton; **13** David Murphy;
14-16 All David Ashby; **17** All Sarah Cuttle; **18-19** All Sarah
Cuttle; **20-21** All Sarah Cuttle; **22-25** All David Murphy; **26-27** All
Mark Winwood; **28-40** All Sarah Cuttle; **41** David Murphy;
43 David Murphy **44** (1) B&Q, (2) Tsunami, **45** (1) Keuco, (2)
Jacuzzi (Jacuzzi UK), (3) Twyford Bathrooms; **46-47** (1) Amtico,
(2) Villeroy & Boch, (3) Jacuzzi (Jacuzzi UK) /David Murphy,
(4) Keuco, (5) Jacuzzi (Jacuzzi UK), (6) Novolon (Forbo), **48-49**
(1) Bathstore.com, (2) Keuco, (3) Bathstore.com, (4) Laufen,
(5) Ikea, (6) Ambiance Bain, (7) Ambiance Bain (8) Villeroy &
Boch; **50-51** (1) Fordham (Jacuzzi UK), (2) Aquaplus, (3) The
Metropolitan Shower Co, (4) Daryl Showering, (5) The
Metropolitan Shower Co, (6) Hansgrohe, (7) Grohe; **52-53** (1)
David Murphy, (2) Lefroy Brooks, (3-4) Jacuzzi (Jacuzzi UK),
(5) Ambiance Bain; (6) Niagara (Jacuzzi UK), (7) Jacuzzi (Jacuzzi
UK), (8) Niagara (Jacuzzi UK), (9) Niagara (Jacuzzi UK), (10) Mike
Newton (11) Grohe; **54-55** (1) Amtico, (2) Marmoleum (Forbo),
(3) Cushionflor (Forbo), (4) Amtico, (5-6) Johnson Tiles; **56-57** (1)
Twyford Bathrooms, (2) Johnson Tiles, (3-4) Twyford Bathrooms,
(5) Johnson Tiles, (5) Sarah Cuttle; **58-59** (1) Jacuzzi (Jacuzzi UK),
(2) Ikea, (3) Lefroy Brooks (4) Myson, (5) Keuco, (6) Schneider,
(7) David Murphy, (8) Villeroy & Boch, (9) Ambiance Bain,
(10) Fordham (Jacuzzi UK); **60** B&Q; **61** David Murphy; **62-63** (1)
The Metropolitan Shower Co, (2) Myson, (3) Niagara (Jacuzzi UK),
(4) Jacuzzi (Jacuzzi UK); **64** All David Ashby; **65** David Murphy;
66-78 All David murphy; **79-81** All Sarah Cuttle; **82-83** (1)
Niagara (Jacuzzi UK), (top) Triton/David Ashby, (2) David Murphy;
84-86 all Sarah Cuttle; **87** All David Murphy; **88** (1) Manhattan
Showers/David Murphy, (2) MK Electric/David Murphy, (3-7) Triton/
David Murphy; **90** (1) Manhattan Showers /Hansgrohe /David
Murphy, (2-3) Daryl Showering/David Murphy, (4) Jacuzzi (Jacuzzi
UK)/Hansgrohe/ David Murphy; **91** David Murphy; **92- 100** all
David Murphy; **101-104** All Sarah Cuttle; **105** David Murphy;
106-107 All Sarah Cuttle/Colchester Tiles; **108-109** (1) Johnson
Tiles, (2) Dalsouple/David Murphy; **110-111** All David Murphy;
112 (1-4) Johnson Tiles/David Murphy, (6-7) Devi/David Murphy,
(8-10) B&Q/David Murphy, **114-118** All Sarah Cuttle; **119** All Sarah
Cuttle except (5) Ambiance Bain; **120-121** (1) David Murphy, (2)
Ikea, (3) Schneider, (4-6) David Murphy; **128** David Murphy

acknowledgements

TOMMY'S
*An enormous thanks to my wife Marie for having to type up every
word of my often illegible scribble onto the PC.*
*Georgie Bennett, Richard Foy, Micky Cunningham and Jimmy the
Joiner for vital information!*
*To Ruth, Angela, Amanda, David, Hannah, Sarah, Anthony, Neal,
Sarah C. and Andres for their help and patience in
co-ordinating and stitching together this book!*
*And a special thanks to my old friends Guy and Sarah at B&Q for
all their help supplying goods for this book.*
*An enormous help was given by Faye at Bathstore.com in
Hammersmith supplying goods for the shoots and a showroom
for photography*

FOR AIREDALE PUBLISHING
We would like to thank the following companies for providing
tools and equipment for this book:
FOR EVERYTHING FROM A–Z
Guy Burtenshaw, Sarah Stonebanks at B&Q
and all the very patient staff at B&Q Yeading
FOR BATHROOMS AND EQUIPMENT
*Faye at Bathstore.com for enormous help supplying endless
bathroom goods and accessories*
Julie Doyle at Crown Paint
Clancy Walker at Propaganda for help with Daryl Showering
Tracy Bates at Hansgrohe for supplying taps and showers
Leslie Holdsworth at Jacuzzi UK for Niagara showers
Andrew Adams at Johnson Tiles for all the tiles used
*Tom Healy at Manhattan Showers for shower enclosures
and Philippa and Melissa at Publicity Engineers.*
Nadia Guen at Myson for the towel warmers
Tanya Johnson for the Triton Showers
Twyford Bathrooms for help with sinks and baths
Julie Mellor at Dalsouple for rubber floor tiles
Neil Patel at Bathwise for using his showroom for photography
Trevor Culpin at Screwfix Direct for tools
Suzanne Mills at Atlas Copco for tools
Nick Cook at DeWalt for tools
Stuart Elsom at Rollins Group for tools